Mycotoxins and Mycotoxicoses

Karuna Singh • Ankita Kumari

Mycotoxins and Mycotoxicoses

Springer

Karuna Singh
Animal Mycology Lab,
Department of Zoology, Mahila
Mahavidyalaya
Banaras Hindu University
Varanasi, Uttar Pradesh, India

Ankita Kumari
Animal Mycology Lab,
Department of Zoology, Mahila
Mahavidyalaya
Banaras Hindu University
Varanasi, Uttar Pradesh, India

ISBN 978-981-19-2369-2 ISBN 978-981-19-2370-8 (eBook)
https://doi.org/10.1007/978-981-19-2370-8

This Springer imprint is published by the registered company Springer Nature Singapore Pte Ltd.
The registered company address is: 152 Beach Road, #21-01/04 Gateway East, Singapore 189721, Singapore

Dedicated to our Parents

Preface

Filamentous fungi such as *Aspergillus*, *Fusarium*, *Penicillium*, and *Alternaria* are common contaminants of food, feed, and indoor environment. During unfavourable conditions, they produce toxic secondary metabolites termed mycotoxins. These mycotoxigenic moulds are producers of both "traditional mycotoxins" such as aflatoxins, ochratoxins, fumonisins, trichothecenes, zearalenone, alternariol monomethyl ether, alternariol, and patulin and "emerging mycotoxins" such as penicillic acid, cyclopiazonic acid, gliotoxin, sterigmatocystin, tenuazonic acid, beauvericin, enniatin, fusaproliferin, moniliformin, culmorin, fusaric acid, sterigmatocystin, butenolide, and emodin.

Mushrooms mostly belong to the subdivision Basidiomycotina; however, few of them are included in Ascomycotina. They are healthy food options, but all of them are not edible. Consumption of toxic species of mushrooms may lead to poisoning syndrome, affecting mainly the gastrointestinal tract and the nervous system. Gyromitrin, amanitin, muscarine, isoxazole derivatives and indole derivatives are well-documented mushroom toxins while others like allenic norleucine are not.

Aflatoxins are major contaminants of corn and corn by-products, peanuts, cotton, pistachios, hazelnuts, Brazil nuts, almonds, and dried figs whilst fumonisins are mostly found in maize but are also prevalent in sorghum, wheat, millet and their products, complete and complementary feeding materials for fishes, hens, calves, pigs, rabbits, horses, lambs, ruminants, and pet animals. *Alternaria* toxins AOH and AME have been found to contaminate wheat, sorghum, barley, sunflower seeds, oats, tomatoes, olives, melons, pecans, and peppers. Ochratoxins produced by *Aspergillus* spp., *Penicillium verrucosum*, and *Fusarium* spp. are general contaminants of wheat, barley, oats, rye, coffee beans, and other plant products including beer wines, spices, and chocolate and feed of pigs, poultry birds, and dairy animals. Zearalenone is a naturally occurring mycotoxin of crops, especially maize infected by *Fusarium* spp. ZEA and trichothecenes persist in the products of wheat, rice, maize, barley, oats, soybeans, and sorghum. A high level of patulin occurs in cereals such as barley, corn, wheat and their processed products, fruits like apple and its products, bananas, grapes, pears, strawberries, apricots, cherries, blueberries, peaches, and plums.

Many fungal metabolites with suspected or known toxicity are designated "emerging mycotoxins". They are known to affect vegetables, cereals and cereal products (bread, beer, and baby food), fruits, processed fruit products, oilseeds, dried fish, naturally fermented sausages, foodstuffs (cheese, miso, sake, and soy sauce), and animal feed. The emerging mycotoxins having less toxic precursors are called T "masked toxins". This category of toxins usually remains undetected during analysis. Despite high incidences of toxicity, they are not legislatively regulated in the food and feed.

Direct or indirect exposure to mycotoxins leads to mycotoxicoses in humans and animals. Aflatoxins are potent carcinogens. Aflatoxicosis together with hepatitis B virus is accountable for thousands of human deaths per year, especially in tropical regions. In chickens, pigs, and cattle, effects of aflatoxins include liver and kidney damage. Ochratoxin A, however, is responsible for urinary tract cancer and kidney damage. Fumonisins cause oesophageal cancer and DNA damage in humans and birth defects in mice. Trichothecenes are highly immunosuppressive while zearalenone has oestrogenic effects in animals and men. Patulin, however, is geno-, hepato-, and nephrotoxic. The inhalation of aflatoxins, ochratoxins, stachybotryotoxins, and *Fusarium* mycotoxins has ill health effects in the occupants of damp indoor environments and mouldy buildings. A cocktail of mycotoxins and indoor contaminants (like smoke) induce respiratory disorders and general intoxication as well. The additive effect of mycotoxins is more hazardous than that of a single one and often leads to more severe co-occurrence poisoning.

Although overlapping symptoms make the diagnosis of mycotoxicoses difficult, some specific clinical signs like weight loss and feed refusal in animals, tremors, hepatic and renal necroses, and lung scarring can be helpful in the differential diagnosis of mycotoxicological effects in humans and animals.

A single method for the control of mycotoxins would not be suitable for all agricultural commodities owing to their different chemical structures. Since very low concentrations of mycotoxins are responsible for toxicity, sensitive and reliable techniques are required for detection. TLC, GC, HPLC, LC-MS, and LC-MS/MS have been established as confirmatory techniques in mycotoxin analysis. Apart from this, ELISA dipstick assay, lateral flow tests NIR, MIR using infrared spectroscopy, molecularly imprinted polymers (MIPs), capillary electrophoresis, fluorescence polarization, labelled optical-read dipstick assays, biosensors based on surface plasmon resonance (SPRs) or fibre-optic probes, and immunological array and nanomaterial-based biosensors are rapid screening methods for detecting the presence of a single or multiple mycotoxins.

The easiest strategy to prevent the formation of mycotoxins is to reduce the growth of moulds. Fertilizers, opportune crop rotation tillage, date of the plantation, selection of genetically modified crops, control of infestation by biological or chemical treatments, removal of the crop, and control of insect and weed are widely discussed aspects to prevent fungal infestation and, thus, mycotoxin production in the fields. Since the discovery of aflatoxins in the 1960s, regulations regarding the intake of mycotoxins have been implemented by several countries. Generally, the establishment of mycotoxin limits is based on factors such as mycotoxin level in the articles, food consumption and toxicological data, food security issues, and methods of analysis. Due to multi-mycotoxin contamination, the complete destruction of mycotoxins is not possible. Therefore, hazard analysis and critical control point (HACCP) based approach is a control programme for processed foods and feeds including strategies for prevention control, good manufacturing practices, and quality control at all stages of production from field to the final consumer. Uses of additives such as propionic acid, ammonia, and silage additives of microbial and enzymatic origin can turn out to be helpful in controlling mould growth and, consequently, mycotoxin production.

Toxicodynamics and toxicokinetics are two important aspects of mycotoxicological studies explaining the relationship between a toxicant and its biological targets. The kinetics of a mycotoxin can be determined by its route/s of entry and ADMET (Absorption, Distribution, Metabolism, Elimination, and Toxicity) properties. Nonetheless, toxicodynamic methods evaluate the mechanism of action and toxicological effects of the toxicant. In silico methods for mycotoxin risk assessment have also been used to obtain supporting data.

This book has ten chapters. The first chapter covers the historical perspective of mycotoxins along with the timeline while the second one provides an overview including the classification of mycotoxins and mycotoxicoses. Comprehensive information on traditional, emerging, and mushroom mycotoxins is given in Chaps. 3, 4, and 5, respectively. Chapter 6 deals with mycotoxins co-occurrence poisoning whereas new and masked mycotoxins are described in Chap. 7. Mycotoxin detection methods are explained in Chap. 8. The ninth chapter includes mycotoxin management strategies. The last (tenth) chapter of the book covers recent developments in mycotoxicokinetics and mycotoxicodynamics.

The book addresses a wide range of readers including mycologists, clinicians, agricultural scientists, chemists, veterinarians, environmentalists, and food scientists, providing all-around knowledge about mycotoxins to date in a simple manner.

Varanasi, Uttar Pradesh, India

Karuna Singh
Ankita Kumari

Contents

Abbreviations

^{1}H NMR	Proton nuclear magnetic resonance
^{3}H-FX	Radioactive Fusarenon-X
3-NPA	3-Nitropropionic acid
AAL toxins	*Alternaria alternata* toxins
AAL-TA	*Alternaria alternata* toxins TA
AAL-TB	*Alternaria alternata* toxins TA
ACROs	Acromelic acids
Af	Aflatoxin
aflJ	Aflatoxin biosynthetic gene
aflR	Regulatory gene *aflR*
AHDA	Amino hexadienoic acid
ALARA	As low as reasonably achievable
ALTCH	Alteichin
ALTs	Altertoxins
AME	Alternariol monomethyl ether
AMV RT	Avian myeloblastosis virus (AMV) reverse transcriptase
AOH	Alternariol
AQA	Analytical quality assurance
ATA	Alimentary toxic aleukia
ATX-I	Altertoxins I
ATX-II	Altertoxins II
ATX-III	Altertoxins III
ATXs	Altertoxins
AUC_{0-t}	Area under the concentration-time curve from time zero to the last measurable concentration
a_w	Water activity
BBB	Blood-brain barrier
BEA	Beauvericin
BUT	Butenolide
bw	Body weight

C-4	Acetyl residues
CaCo-2	Colorectal adenocarcinoma cells-2
CCPs	Critical control points
CEC	Commission of the European Communities
CFCF	Cell-free culture filtrate
CIT	Citrinin
Cl	Clearance
Cmax	Maximum (or peak) serum concentration
CNS	Central nervous system
CPA	Cyclopiazonic acid
CT	Computed tomography
CTN	Citrinin
CUL	Culmorin
CYP 450	Cytochromes P450
CYP1A2	Cytochrome P450 family 1 subfamily A member 2
CYP3A4	Cytochrome P450 family 3 subfamily A member 4
CYP3A5	Cytochrome P450 family 3 subfamily A member 5
D3G	Deoxynivalenol-3-glucoside
DAS	Diacetoxyscirpenol
DDGS	Dried distillers' grains with soluble
DESI	Desorption electrospray ionization
DMSO	Dimethyl sulfoxide
DNA	Deoxyribonucleic acid
DOM	Deepoxy-DON
DOM-1	Deepoxy metabolite of DON
DON	Deoxynivalenol
EA	Ergot alkaloids
EC	Oesophageal cancer
EFSA	European Food Safety Authority
ELEM	Equine leukoencephalomalacia
ELISA	Enzyme-linked immunosorbent assay
EMO	Emodin
ENN	Enniatins
ENNs	Enniatins
EU	European Union
F	Bioavailability
F. verticillioides	MRC 826
FA	Fusaric acid
FAO	Food and Agriculture Organization of the United Nations
FAPAS	Food Analysis Performance Assessment Scheme
FB	Fumonisins B
FB1	Fumonisin B_1
FLD	Fluorescence detection
FMEA	Failure mode and effect analysis

FPIA	Fluorescence polarization immunoassay
FTIR	Fourier transform infrared spectroscopy
Fum5 gene	Fumonisin synthase gene
FUS	Fusaproliferin
FX	Fusarenon-X
GABA	Gamma-aminobutyric acid
GAP	Good Agricultural Practices
GC	Gas chromatography
GC-MS	Gas chromatography-mass spectrometry
GDP	Guanosine diphosphate
GHP	Good Hygienic Practices
GI	Gastrointestinal tract
GMP	Good Manufacturing Practices
GSP	Good Storage Practices
GT	Gliotoxin
HACCP	Hazard analysis and critical control point
HPLC	High performance liquid chromatography
IARC	International Agency for Research on Cancer
IBO	Ibotenic acid
IDH	Isoepoxydon dehydrogenase
IEC	Intestinal epithelial cell
IG	Intra-gastric
IP	Intra-peritoneally
iv	Intra-venous
JECFA	Joint FAO/WHO Expert Committee on Food Additives
KDa	Kilodalton
LADME	Liberation, absorption, distribution, metabolism and excretion
LC	Liquid chromatography
LC-MS	Liquid chromatography-mass spectrometry
LC-MS/MS	Liquid chromatography-mass spectrometry/mass spectrometry
LFD	Lateral flow devices
LLC-PK1 renal cells	Epithelial cell line originally derived from porcine kidneys
LOD	Limit of detection
LOQ	Limit of quantification
LSD	Lysergic acid diethylamide
MFH	*N*-methyl-*N*-formylhydrazine
MH	Monomethylhydrazine
MON	Moniliformin
MPA	Mycophenolic acid
MRT	Marginal rate of transformation
MS	Mass spectrometry
MUS	Muscimol
MW	Molecular weight

Na-K ATPase	Sodium potassium pump adenosine triphosphatase enzyme
NASBA	Nucleic acid sequence-based amplification
NIH/3T3	3-day transfer, inoculum 3 × 10^5 cells from NIH mouse embryonic fibroblast cells
NIR	Near-infrared
NIV	Nivalenol
nor-1 gene	Neuron-derived orphan receptor-1
NRPSs	Non-ribosomal peptide synthetases
OTA	Ochratoxin A
OTB	Ochratoxin B
OT-GSH	OTA-glutathione conjugate
OTHQ	OTA-hydroquinone
OTQ	OTA-quinone
PA	Penicillic acid
PAT	Patulin
PCR	Polymerase chain reaction
PIDI	Provisional tolerable daily intake
PIWI	Provisional tolerable weekly intake
pks-A	Polyketide synthases
po	*Per os*
PPE	Porcine pulmonary oedema
ppm	Parts per million
RNA	Ribonucleic acid
Rnase H	Ribonuclease H
RP-HPLC	Reversed phase high performance liquid chromatography
RT PCR	Real-time polymerase chain reaction
RT	Room temperature
SH	Sulfhydryl groups
SLUDGE	Salivation, lacrimation, urination, defecation, gastrointestinal distress, and emesis
SPE Cartridges	Solid phase extraction cartridge
SPSR	Surface plasmon resonance sensors
SSR	Self-sustained sequence replication
STC	Sterigmatocystin
STE III	Stemphyltoxins III
STE	Stemphyltoxins
$t_{1/2}$	Half-life
T-2	T-2 toxin
T7 RNA	T7 bacteriophage that catalyses the formation of RNA from DNA
T7 RNA polymerase	RNA polymerase from the T7 bacteriophage
TCs	Trichothecenes
TDI	(Maximum) tolerable daily intake
TeA	Tenuazonic acid

TEER	Transepithelial electrical resistance
TEN	Tentoxin
TLC	Thin layer chromatography
TP53	Tumour protein *p53*
Tri5	Trichodiene synthase gene
UHPLC	Ultra-high-performance liquid chromatography
UV	Ultraviolet
V_d	Volume of distribution
WHO	World Health Organization
Z14G	Zearalenone-14-β-D-glucopyranoside
Z14S	Zearalenone-14-sulphate
ZEN	Zearalenone
α-ZEL	α-zearalenol
β-ZEL	β-zearalenol

Chapter 1
Historical Aspect

1.1 Introduction

"Wailing and writhing men collapsed in the street: others fell over and foamed in epileptic fits whilst some vomited and showed signs of insanity. Many of them shouted 'Fire! I'm burning'. It was an invisible fire that separated the flesh from the bones and consumed it. Men, women and children died in unbearable agonising pain".

The above words were quoted by a chronicler in the tenth century to document a disease that affected several places in Europe in 943 AD. The victims experienced a burning sensation, and many of them visited the shrine of St Anthony in France with the faith of being cured. Hence, the disease came to be known as "St Anthony's fire". St Anthony's fire, now known as ergotism, was caused due to the intake of rye infested with the mould *Claviceps purpurea* that produced "ergot alkaloids". It became an epidemic in Europe in the tenth century (Manual on the Application of the HACCP System in Mycotoxin Prevention and Control 2001). This human illness caused by a fungus has been known for a long time, probably by the Greeks and Romans. In the middle Ages, ergotism was a well-known disease that spread not only in France but throughout Europe, the United Kingdom, Scandinavia, and Russia. Outbreaks of fatal ergotism were repeatedly reported from 900 to 1800 AD in central Europe, causing hundreds and thousands of deaths. However, this epidemic is sometimes confused with bubonic plague (Barger 1931; Pitt and Miller 2017).

The idea of fungal toxins and their effects have always been ignored. Mycotoxins have existed ever since the practice of cultivating crops and storing them for future use has come into existence. As stated by Pitt and Miller (2017), mycotoxins have always been "black swans" as their effects are not specific, and it is difficult to trace the connection between mycotoxins and illnesses they cause (Pitt and Miller 2017).

K. Singh, A. Kumari, *Mycotoxins and Mycotoxicoses*,
https://doi.org/10.1007/978-981-19-2370-8_1

1.2 History of Mycotoxins

1.2.1 Ergotism

Mycotoxins have a long history in human diseases, and ergotism is believed to be the oldest recognized mycotoxicosis. In the middle Ages, the European literature has numerous mysterious reports of outbreaks (From Poisoning to Pharmacy: A Tale of Two Ergots). Rye was a staple food for the farmers in Europe. *Claviceps purpurea* is an ergot fungus that infests rye and forms a solid biomass on its developing seeds. The seeds when harvested and ground produced a poisonous cocktail of toxins that caused convulsions, gangrene, and spontaneous abortions (Matossian 1989; http://www.botany.hawaii.edu/faculty/wong/BOT135/LECT11.HTM).

In the nineteenth century, the structure of ergot alkaloids, lysergic acid derivatives (LSD), was first elucidated (Barger 1937). Some ergot alkaloids are now used in pharmaceuticals after knowing their mechanism of action. In 1954, the last outbreak of ergotism in Europe was reported in the French village of Pont St Esprit (Bread of madness 1951). It affected 200 people and led to the death of four amongst them from cardiovascular collapse (Fuller 1968; Gabbai 1951). Another incidence of occupational exposures to ergot during milling was reported recently that led to severe toxicity to the millers (Stange et al. 1998). Ergot has become quite common in places of Europe and North America in the last decade (Krska and Crews 2008; Menzies and Turkington 2015).

The genus *Claviceps* forms a sclerotium that is commonly known as ergot, which produces ergot alkaloids. Following infection, the seed or kernel of the plant is replaced by the sclerotium, which is a dark-coloured, hard fungal mass. Other genera producing ergot alkaloids are *Rhizopus*, *Penicillium*, and *Aspergillus* spp. (Passmore 1951). *Claviceps* spp. are known to yield over 70 different types of alkaloids (Schiff 2006). Ergot alkaloids can be chemically divided into derivatives of the following (Van Dongen and de Groot 1995):

1. Isolysergic acid (e.g. ergotaminine)
2. Lysergic acid (e.g. ergotamine and ergocristine)
3. Dimethylergoline (clavines, e.g. agroclavine)

The type of alkaloid present and the type of ergotism caused largely depend on the source of the alkaloid (Forgacs 1962). Ergotamine-ergocristine alkaloids synthesized by *Claviceps purpurea* cause vasoconstriction leading to a gangrenous form of ergotism. The primary symptoms included oedema of the legs accompanied by severe pain. Paraesthesias followed by gangrene at the tendons and painless demarcations are also observed. The last outbreak of gangrenous ergotism was recorded in Ethiopia in 1977–1978 when 140 people were affected and the mortality recorded was 34% (Stoloff 1976). The second kind of ergotism is the convulsive form that results from the consumption of clavine alkaloids synthesized by *Claviceps fusiformis* and was last recorded in India in 1975, affecting 78 people (Bennett and Klich 2003; Pitt 1996; Zain 2011). The incubation period for the toxin is 1–48 h. It was characterized by gastrointestinal disturbances including giddiness, nausea, and vomiting, further affecting the central nervous system and showing symptoms such as prolonged sleepiness, drowsiness, blindness, convulsions, twitching, and paralysis, but no deaths were recorded. Ergotism is also known to cause reduced fertility as sublethal exposure of ergot alkaloids showed reproductive toxicity (Barger 1931; Matossian 1981).

The cleaning and milling processes nowadays remove most of the alkaloids and the flour, and if it contains any, it is present at a very low concentration; therefore, cases of ergotism are extremely rare now. Moreover, the alkaloids are heat labile and destroyed during cooking and baking (Pitt and Miller 2017).

1.2.2 Alimentary Toxic Aleukia (ATA)

A human illness called "taumelge treide", also known as staggering grains, was suspected to be caused by trichothecenes (TCs) and was observed for the first time in the 1980s in Siberia. Symptoms of the illness included headache, vomiting, and vertigo (Pestka and Smolinski 2005). Cereal grains contaminated with *Fusarium sporitrichoides* and *Fusarium poae* were implicated in alimentary toxic aleukia in Russia from 1932 to 1947 (Zain 2011). Although it was reported in the nineteenth century, a severe outbreak took place during the World War II when 100,000 people died in the Orenburg region in USSR (Marasas 2001; Yazar and Omurtag 2008). Symptoms included oesophageal pain, mucous membrane hyperaemia, gastroenteritis, laryngitis, asphyxiation, and vertigo.

Misuse of TCs as a potential biological or chemical agent in wars was also a reason behind its research. In the late 1970s and early 1980s, TCs were proclaimed to be used on civilian and refugee populations in Afghanistan, Laos, and Kampuchea as weaponized aerosols. Symptoms such as bleeding, dizziness, dyspnoea, fever, nausea, and vertigo were reported. These events were called "yellow rain", as described by the witnesses, but the claim that TCs were accountable for the reported symptoms was controversial (Sudakin 2003). TCs were found to be positively correlated to the occurrence of oesophageal cancer in humans (Hsia et al. 1988).

1.2.3 Mouldy Corn Disease

Prior to 1900 in Italy, researchers believed that consumption of mouldy corn led to the development of an illness in children. In the early 1950s, an epidemic of mouldy corn disease in the south-eastern United States occurred where hundreds of wild pigs foraging in cultivated corn fields became ill and many died (Elshafie et al. 2011; Hunter 1989; Kendrick 1992; Matossian 1989) (http://www.botany.hawaii.edu/faculty/wong/BOT135/LECT11.HTM).

1.2.4 Turkey-X Disease

In 1960 in England, almost 100,000 turkeys and some other domestic birds died due to a syndrome that was named "turkey-X disease". Investigators isolated *Aspergillus flavus* from the peanut meal fed to the birds. The isolated fungus was further inoculated in the feed given to the turkeys. The turkeys died within a short span of time after they were fed with the inoculated feed. The signs, symptoms, and internal lesions observed in the turkeys matched that of the birds that died earlier, thereby proving that the causative agent of the disease was *Aspergillus flavus* (Richard 2008) (http://www.botany.hawaii.edu/faculty/wong/BOT135/LECT11.HTM).

1.2.5 Leukoencephalomalacia (ELEM), Porcine Pulmonary Oedema (PPE), and Human Oesophageal Cancer (EC)

In 1970 in South Africa, an outbreak known as equine leukoencephalomalacia (ELEM) occurred in the fields. The disease is also known as the hole in the head syndrome due to its characteristic feature of showing liquefactive necrotic lesions in the white matter of the cerebral hemispheres of horses. *F. verticillioides* (*F. moniliforme*) was the main fungus isolated from mouldy corn associated with the outbreak (Marasas 2001). Shortly thereafter, in 1989, high levels of fumonisins in the United States corn crop led to large-scale field outbreaks of ELEM and PPE (porcine pulmonary oedema) in horses and pigs, respectively, in the United States (Sudakin 2003). Subsequently, in the Transkei region of South Africa, studies showed the possible role of fungal toxins in the aetiology of human oesophageal cancer (EC). In the southern part of Transkei, the incidence rate of EC was among the highest in the world, whereas the rate was low in the northern part. Home-grown corn was the staple diet in both areas, and the most prevalent fungus in the corn was *F. verticillioides* in areas with a high incidence of EC (Marin et al. 2013).

1.2.6 Mouldy Sugarcane Poisoning

During winters (February and March) of the 1970s and 1980s, "mouldy sugarcane poisoning", an acute form of food poisoning, occurred in northern China in 13 provinces. *Arthrinium* sp. was the cause of poisoning as it contaminated sugarcane that was stored for 2 months or more. In outbreaks during the period 1972 to 1988, almost 884 people got affected, with 10% fatalities (Peraica et al. 1999).

1.2.7 Cardiac Beriberi

Acute cardiac beriberi is another human mycotoxicosis of significance reported in Japan, especially in the second half of the last century. Symptoms of the disease included breathing difficulties, nausea, and vomiting. Severe pain and distress were followed after 2–3 days. It caused progressive paralysis that even led to failure of the respiratory system and eventually death (Pitt 1989).

Consumption of polished rice results in vitamin deficiencies, and the condition is known as beriberi. Studies carried out by Uraguchi in 1971 showed that acute cardiac beriberi might be a toxicosis rather than a vitamin deficiency (Uraguchi 1967). The prevalence of acute cardiac beriberi suddenly decreased in Japan in 1910. According to the observations of Uraguchi, the reduction in the sale of mouldy rice by the implementation of government laws also led to a decrease in the cases of cardiac beriberi. However, beriberi caused due to the polished rice consumption remained unaltered. It was also observed that acute cardiac beriberi often affected young healthy individuals. *Penicillium citreonigrum*, though a rare species that produced a mycotoxin citreoviridin, was isolated from the rice of that region. Therefore, he concluded that cardiac beriberi was a mycotoxicosis caused by the consumption of citreoviridin and not by the consumption of polished rice. The disease does not prevail in Japan anymore but might exist in some parts of the continent (Pitt 1989).

1.2.8 Onyalai

Onyalai was a disease endemic to Africa, especially affecting the Southern Saharan regions for over a period of 80 years. It was found to affect the rural population more than the urban as most of the rural people were dependent on millet as a staple diet. Rabie et al. (1975) proposed that the illness might be related to a mycotoxin. *Phoma sorghina*, a toxigenic fungus, was isolated from millet consumed by people affected. Most of the symptoms of onyalai were reproduced in rats fed with maize and wheat contaminated with *P. sorghina* (Pitt 1989).

1.2.9 Aflatoxin Poisoning in Dogs

In 2020 and 2021, pet food manufactured by *Midwestern Pet Foods recalled its pet food as* around 70 dogs died from aflatoxin poisoning from December 2020 to January 12, 2021. *The U.S. Food and Drug Administration investigated the pet food product manufactured by the company and found out that it contained potentially fatal levels of aflatoxins. The product was recalled from the countries to which the pet food was exported* (FDA Alert 2021; Tyko 2021).

1.2.10 Other Incidences

In the past years, disease outbreaks of mycotoxins have been reported, such as vascular ergotism in Ethiopia, enteric ergotism in India, deoxynivalenol mycotoxicosis in India and China, and aflatoxic hepatitis in India and Kenya. Involvements of staple foods like corn, wheat, or pearl millet and unseasonable drought or rains either during the growing season or harvest were common to all these outbreaks (Bhat and Miller 1991).

The following timeline shows the outbreaks and incidences of mycotoxicoses that occurred during the course of time in different parts of the world (Fig. 1.1).

1.3 Discovery and Synthesis of Mycotoxins

1.3.1 Cyclopiazonic Acid (CPA)

Holzapfel in 1968 discovered and chemically characterized CPA. He extracted CPA from cultures of *Penicillium cyclopium* Westling (strain CSIR 1082) (correct name: *Penicillium griseofulvum* Dierckx) as a major toxin. It was originally discovered while routine toxicity screening of *P. cyclopium* isolated from groundnuts that were connected with acute toxicity in rats and ducklings were being investigated (Ostry et al. 2018).

1.3.2 Ochratoxin A (OTA)

Chemical characterization and isolation of OTA were first performed in the year 1965 in South Africa (Hussein and Brasel 2001; Akande et al. 2006). When *Aspergillus ochraceus* was intentionally inoculated to corn meal, OTA was isolated as a toxic metabolite (Malir et al. 2016).

1.3.3 Trichothecenes

Freeman and Morrison in the year 1948 first isolated the trichothecene 'trichothecin' from *Trichotheceum roseum* (Pitt and Miller 2017). In the year 1961, Brian et al. characterized diacetoxyscirpenol (DAS) from *Fusarium equiseti* first and then nivalenol (NIV) and T-2 toxin from *F. sporotrichioides*, followed by nivalenol (NIV) and T-2 toxin, from *F. sporotrichioides*. The discovery of *Fusarium* mycotoxins took a turn and led to the discovery of trichothecenes from other genera in 1980 when 4-deoxynivalenol (DON) was being isolated from wheat in Eastern

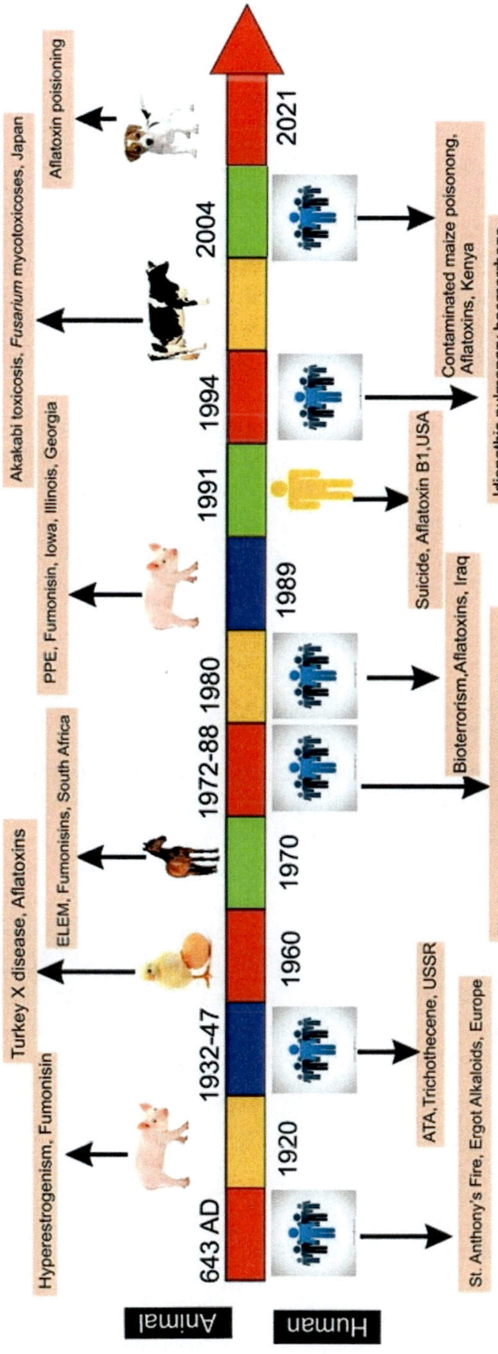

Fig. 1.1 Outbreaks of mycotoxicoses that occurred during the course of time in different parts of the world from 643 AD to 2021

North America (Brian et al. 1961; Tatsuno et al. 1968; Bamburg et al. 1968; Marasas et al. 1984; Hsia et al. 1988; Shank et al. 2011).

1.3.4 Fumonisins

Fumonisins were first discovered in 1988 in South Africa. Cultures of *F. verticillioides* strain MRC 826 isolated from mouldy corn in Transkei produced toxic substances that caused ELEM in horses, PPE syndrome in pigs, and hepatic malignancy in rats. The livers of rats were used to carry out a short-term cancer initiation/promotion assay to isolate the cancer-causing substance/s from the culture material. Two carcinogenic mycotoxins were isolated and discovered from cultures of *F. verticillioides* MRC 826, which were fumonisins B1 and B2, at a programme being carried out on Mycotoxins and Experimental Carcinogenesis of the Medical Research Council in Tygerberg in South Africa. The chemical structure of the mycotoxins was further elucidated, and it was proved that the mouldy corns contained these toxins. Soon after its discovery, large-scale field outbreaks of ELEM and PPE occurred in the United States, followed by EC in Transkei and China (Marasas 2001).

1.3.5 Patulin

The discovery of patulin took place simultaneously by several groups. Therefore, it came to be known by different names such as expansine when discovered by Van Luijk (1938), claviformin by Chain et al. (1942), clavacin by Anslow et al. (1943), clavatin by Bergel et al. (1943), gigantic acid by Philpot (1943), and myocin C by DeRosnay et al. (1952). However, it was first isolated in the 1940s as a water-soluble lactone by Stott and Bullerman (1975). In the year 1976, a broad-spectrum anti-fungal antibiotic nature of patulin was reported by Korzybski as it inhibited the growth of more than 75 different Gram-positive and Gram-negative bacteria (Ciegler et al. 1976). Later studies proved that patulin was not only toxic to bacteria and other fungi but also to plants and animals (Moake et al. 2005).

1.3.6 Penicillic Acid

Alsberg and Black were the first to isolate penicillic acid from *Penicillium puberulum* Bainier. Later on, Birkinshaw, Oxford, and Raistrick isolated it from *Penicillium cyclopium* Westling and discovered its structure (IV). The first practical synthesis of penicillic acid was reported by Ralph Raphael in 1947a, who worked with penicillin during World War II. It was then found to have inhibitory activity against bacteria, and its activity was more than penicillin against Gram-negative organisms (Raphael 1947b).

1.3.7 Gliotoxin

Isolation of cultures from wood fungus *Gliocladium fimbriatum* led to the discovery of gliotoxin by Weindling and Emerson in 1936. It was later isolated from other microfungi such as *Penicillium terlikowskii and Aspergillus fumigatus* (Bennett and Klich 2003; Fukuyama et al. 1981).

1.3.8 Sterigmatocystin

The discovery of sterigmatocystin took place when the investigation of the metabolites of *Aspergillus versicolor* (Vuill.) Tiraboschi was being carried out (Davies et al. 1960; Roberts and Underwood 1962; Bullock et al. 1962). Ariatoxin, a toxic metabolite of *Aspergillus flavus* Link, was found to be structurally similar to sterigmatocystin. Ariatoxin has a bicyclic furan attached to a coumarin moiety, while sterigmatocystin has a bifuran structure attached to a xanthone nucleus (Asao et al. 1965; Lillehoj and Ciegler 1968).

1.3.9 Tenuazonic Acid

Rosett et al. were the first to isolate tenuazonic acid from culture filtrates of different *Alternaria tenuis* strains. The discovery of tenuazonic acid was made during an antiviral screening programme that used culture extracts of *Aspergillus* to check its cytopathic effect against the measles virus and antibacterial activity. An antiviral compound was isolated that acted against measles and poliovirus, which later came to be known as tenuazonic acid (Miller et al. 1963).

1.3.10 Enniatin A

Fusarium orthoceras var. *Enniatinum* was the first organism reported to produce enniatin A in 1947 by Gaumann et al. Enniatin A showed antibacterial activity against mycobacterium. Soon after that, a compound named lateritiin-I was isolated from *Fusarium lateritium*, which was identical to enniatin A. Also, 27 years later, in 1974, previously reported enniatin A was identified as a mixture of *threo* and *erythro* optical isomers by Audhya and Russel. Along with enniatin A, enniatin B and C were also isolated. To check its ionophoric property, more and more production of enniatin A was performed by scientists, which led to the discovery of ennianitin A-P (Matossian 1989; Sy-Cordero et al. 2012).

1.3.11 Fusaproliferin

In the year 1993, fusaproliferin was isolated for the first time from a strain of *Fusarium proliferatum* and then from *Fusarium subglutinans* (Logrieco et al. 1996; Meca et al. 2009; Randazzo et al. 1993; Ritieni et al. 1995; Shephard et al. 1999). Later, production of fusaproliferin has also been reported from *Fusarium antophilum*, *Fusarium begoniae*, *Fusarium bulbicola*, *Fusarium circinatum*, *Fusarium concentricum*, *Fusarium succisae*, and *Fusarium udum* (Moretti et al. 2007; Santini et al. 2012).

1.3.12 Moniliformin

Moniliformin was isolated for the first time from *F. moniliforme* associated with corn. It was named after the organism it was isolated from; however, it turned out to be a misnomer as later on it was found that *F. moniliforme* (now called *F. verticillioides*) strains could produce moniliformin in very low quantities (Bianchini and Bullerman 2014).

1.3.13 Fusaric acid

Fusarium species are the primary source of fusaric acid, but *Fusarium heterosporium* was the fungus from which it was isolated for the first time as a picolinic acid derivative and an antibiotic (wilting agent) (Yabuta et al. 1934).

1.3.14 Culmorin

Culmorin was first isolated in 1937 and structurally characterized in the year 1967 (Barton and Werstiuk 1967; Ashley et al. 1937). *F. culmorum* was the organism from which it was isolated; hence, the name culmorin was given. Lately, a basal species of the *F. graminearum* species complex, *F. praegraminearum*, was identified that was able to produce culmorin as an ancestral trait (Weber et al. 2018).

References

Akande KE, Abubakar MM, Adegbola TA, Bogoro SE (2006) Nutritional and health implications of mycotoxins in animal feeds: a review. Pak J Nutr 5(5):398–403

Raphael (1947a) Alsberg and Black

Anonymous (1951) "Bread of madness" infects a town. Life Mag 31(11):25–27

Anslow WK, Raistrick H, Smith C (1943) Anti-fungal substances from moulds. Part I. Patulin (Anhydro-3-hydroxymethylene-tetrahydro-1: 4-pyrone-2-carboxylic acid), A metabolic product of *Penicillium patulum* bainier and *Penicillium expansum* (link) thom. J Soc Chem Indus 62(12):236–238. https://doi.org/10.1002/jctb.5000621208

Asao T, Büchi G, Abdel-Kader MM, Chang SB, Wick EL, Wogan GN (1965) The structures of aflatoxins B and G1. J Am Chem Soc 87(4):882–886. https://doi.org/10.1021/ja01082a031

Ashley JN, Hobbs BC, Raistrick H (1937) Studies in the biochemistry of micro-organisms: the crystalline colouring matters of *Fusarium culmorum* (WG Smith) Sacc. and related forms. Biochem J 31(3):385. https://doi.org/10.1042/bj0310385

Bamburg JR, Riggs NV, Strong FM (1968) The structures of toxins from two strains of *Fusarium tricinctum*. Tetrahedron 24(8):3329–3336. https://doi.org/10.1016/S0040-4020(01)92631-6

Barger G (1931) Ergot and ergotism. A monograph based on the Dohme lectures delivered in Johns Hopkins University, Baltimore, London, Edinburgh. Gurney and Jackson, London

Barger G (1937) The alkaloids of ergot. Analyst 62(734):340–354

Barton DH, Werstiuk NH (1967) The constitution and stereochemistry of culmorin. Chem Commun 1:30–31

Bennett JW, Klich M (2003) Mycotoxins. Clin Microbiol Rev 16:497–516

Bergel F, Morrison AL, Moss AR, Klein R, Rinderknecht H, Ward JL (1943) An antibacterial substance from *Aspergillus clavatus* and *Penicillium claviforme* and its probable identity with patulin. Nature 152(3869):750

Bhat RV, Miller JD (1991) Mycotoxins and food supply. Food, nutrition and agriculture-Food for the future. FAO, Rome

Bianchini A, Bullerman LB (2014) Mycotoxins, classification. In: Encyclopedia of food microbiology. Elsevier, Amsterdam, pp 854–861

Brian PW, Dawkins AW, Grove JF, Hemming HG, Lowe D, Norris GL (1961) Phytotoxic compounds produced by *Fusarium equiseti*. J Exp Bot 12(1):1–2

Bullock E, Roberts JC, Underwood JG (1962) 808. Studies in mycological chemistry. Part XI. The structure of isosterigmatocystin and an amended structure for sterigmatocystin. J Chem Soc:4179–4183

Chain E, Florey HW, Jennings MA (1942) An antibacterial substance produced by *Penicillium claviforme*. Br J Exp Pathol 23(4):202

Ciegler A, Beckwith AC, Jackson LK (1976) Teratogenicity of patulin and patulin adducts formed with cysteine. Appl Environ Microbiol 31:644–667. https://doi.org/10.1128/aem.31.5.664-667.1976

Davies JE, Kirkaldy D, Roberts JC (1960) Studies in mycological chemistry. Part VII. Sterigmatocystin, a metabolite of *Aspergillus versicolor* (Vuillemin) tiraboschi. J Chem Soc 437:2169–2178

DeRosnay CD, Martin-Dupont JR, Jensen R (1952) An antibiotic, myocin. CJ Med Bordeaux et Sud-Ouest 129:189

Elshafie SZ, ElMubarak A, El-Nagerabi SA, Elshafie AE (2011) Aflatoxin B1 contamination of traditionally processed peanuts butter for human consumption in Sudan. Mycopathologia 171(6):435–439. https://doi.org/10.1021/acs.jafc.6b04494

FAO/IAEA Training, Reference Centre for Food and Pesticide Control (2001) Manual on the application of the HACCP system in mycotoxin prevention and control (No. 73). Food & Agriculture Organization, Rome

Forgacs J (1962) Mycotoxicosis. Adv Vet Sci 7:273–383

Fukuyama T, Nakatsuka SI, Kishi Y (1981) Total synthesis of gliotoxin, dehydrogliotoxin and hyalodendrin. Tetrahedron 37(11):2045–2078. https://doi.org/10.1016/S0040-4020(01)97960-8

Fuller JG (1968) The day of St Anthony's fire. McMillan, New York, NY

Gabbai L (1951) Ergot poisoning at Pont St. Esprit. Br Med J 2(4732):650

Hsia CC, Wu JL, Lu XQ, Li YS (1988) Natural occurrence and clastogenic effects of nivalenol, deoxynivalenol, 3-acetyl-deoxynivalenol, 15-acetyl-deoxynivalenol, and zearalenone in corn from a high-risk area of esophageal cancer. Cancer Detect Prev 13(2):79–86

Hunter BT (1989) Mycotoxins: a significant public health problem. Consumers Res 72(6):8–9

Hussein HS, Brasel JM (2001) Toxicity, metabolism, and impact of mycotoxins on humans and animals. Toxicology 167(2):101–134. https://doi.org/10.1016/S0300-483X(01)00471-1

Kendrick B (1992) Mycotoxins in food and feed in the fifth kingdom, vol 2. Mycologue Publications, Waterloo ON, pp 316–331

Krska R, Crews C (2008) Significance, chemistry and determination of ergot alkaloids: a review. Food Addit Contam 25(6):722–731. https://doi.org/10.1080/02652030701765756

Lillehoj EB, Ciegler A (1968) Biological activity of sterigmatocystin. Mycopathol Mycol Appl 35(3):373–376

Logrieco A, Moretti A, Fornelli F, Fogliano V, Ritieni A, Caiaffa MF, Randazzo G, Bottalico A, Macchia L (1996) Fusaproliferin production by Fusarium subglutinans and its toxicity to Artemia salina, SF-9 insect cells, and IARC/LCL 171 human B lymphocytes. Appl Environ Microbiol 62(9):3378–3384. https://doi.org/10.1128/aem.62.9.3378-3384.1996

Malir F, Ostry V, Pfohl-Leszkowicz A, Malir J, Toman J (2016) Ochratoxin A: 50 years of research. Toxins 8(7):191. https://doi.org/10.3390/toxins8070191

Marasas WF (2001) Discovery and occurrence of the fumonisins: a historical perspective. Environ Health Perspect 109(2):239–243. https://doi.org/10.1289/ehp.01109s2239

Marasas WF, Nelson PE, Toussoun TA (1984) Toxigenic *Fusarium* species, identity and mycotoxicology. Pennsylvania State University Press, University Park, PA

Marin S, Ramos AJ, Cano-Sancho G, Sanchis V (2013) Mycotoxins: occurrence, toxicology, and exposure assessment. Food Chem Toxicol 60:218–237. https://doi.org/10.1016/j.fct.2013.07.047

Matossian MK (1981) Mold poisoning: an unrecognized English health problem, 1550–1800. Med Hist 25(1):73–84

Matossian MK (1989) Poisons of the past: molds, epidemics, and history, vol 16. Yale University Press, New Haven, CT, London, p 190

Meca G, Sospedra I, Soriano JM, Ritieni A, Valero MA, Mañes J (2009) Isolation, purification and antibacterial effects of fusaproliferin produced by *Fusarium subglutinans* in submerged culture. Food Chem Toxicol 47(10):2539–2543. https://doi.org/10.1016/j.fct.2009.07.014

Medicine, Center for Veterinary (2021) FDA Alert: certain lots of sportmix pet food recalled for potentially fatal levels of aflatoxin. FDA, Silver Spring, MD

Menzies JG, Turkington TK (2015) An overview of the ergot (*Claviceps purpurea*) issue in western Canada: challenges and solutions. Can J Plant Pathol 37(1):40–45. https://doi.org/10.1080/07060661.2014.986527

Miller F, Rightsel WA, Sloan BJ, Ehrlich J, French JC, Bartz QR, Dixon GJ (1963) Antiviral activity of tenuazonic acid. Nature 200(4913):1338–1339

Moake MM, Padilla-Zakour OI, Worobo RW (2005) Comprehensive review of patulin control methods in foods. Compr Rev Food Sci Food Saf 4(1):821. https://doi.org/10.1111/j.1541-4337.2005.tb00068.x

Moretti A, Mulè G, Ritieni A, Logrieco A (2007) Further data on the production of beauvericin enniatins and fusaproliferin and toxicity to Artemia salina by Fusarium species of *Gibberella fujikuroi* species complex. Int J Food Microbiol 118(2):158–163. https://doi.org/10.1016/j.ijfoodmicro.2007.07.004

Ostry V, Toman J, Grosse Y, Malir F (2018) Cyclopiazonic acid: 50th anniversary of its discovery. World Mycotoxin J 11(1):135–148. https://doi.org/10.3920/WMJ2017.2243

Passmore R (1951) Famine in India. An historical survey. Lancet 2:303–307

Peraica M, Radić B, Lucić A, Pavlović M (1999) Toxic effects of mycotoxins in humans. Bull World Health Organ 77(9):754

Pestka JJ, Smolinski AT (2005) Deoxynivalenol: toxicology and potential effects on humans. J Toxicol Environ Health B 8(1):39–69. https://doi.org/10.1080/10937400590889458

Philpot FJ (1943) A penicillin-like substance from *Aspergillus giganteus* Wehm. Nature 152(3868): 725

Pitt JI (1989) An introduction to mycotoxins. Di dalam. In: Semple RL, Frio AS, Hicks PA, Lozare JV (eds) Mycotoxin prevention and control in food grains. Proceedings of Assistance for The Training Course. Bangkok, p 31

Pitt JI (1996) What are mycotoxins? Aust Mycotox Newsl 7(4):1

Pitt JI, Miller JD (2017) A concise history of mycotoxin research. J Agric Food Chem 65(33): 7021–7033. https://doi.org/10.1021/acs.jafc.6b04494

Rabie CJ, van Rensburg SJ, van der Watt JJ, Lubben A (1975) Onyalai-the possible involvement of a mycotoxin produced by Phoma sorghina in the aetiology. S Afr Med J 49(40):1647–1650. https://hdl.handle.net/10520/AJA20785135_24152

Randazzo G, Fogliano V, Ritieni A, Mannina L, Rossi E, Scarallo A, Segre AL (1993) Proliferin, a new sesterterpene from *Fusarium proliferatum*. Tetrahedron 49(47):10883–10896. https://doi.org/10.1016/S0040-4020(01)80241-6

Raphael RA (1947b) Synthesis of the antibiotic, penicillic acid. Nature 160(4060):261–262

Richard JL (2008) Discovery of aflatoxins and significant historical features. Toxin Rev 27(3–4): 171–201. https://doi.org/10.1080/15569540802462040

Ritieni A, Fogliano V, Randazzo G, Scarallo A, Logrieco A, Moretti A, Manndina L, Bottalico A (1995) Isolation and characterization of fusaproliferin, a new toxic metabolite from *Fusarium proliferatum*. Nat Toxins 3(1):17–20. https://doi.org/10.1002/nt.2620030105

Roberts JC, Underwood JG (1962) Studies in mycological chemistry. Part IX. Synthesis of methyl 1, 3, 8-trimethoxyxanthone-4-carboxylate—a degradation product of sterigmatocystin. J Chem Soc 385:2060–2063

Santini A, Meca G, Uhlig S, Ritieni A (2012) Fusaproliferin, beauvericin and enniatins: occurrence in food–a review. World Mycotoxin J 5(1):71–81. https://doi.org/10.3920/WMJ2011.1331

Schiff PL Jr (2006) Ergot and its alkaloids. Am J Pharm Educ 70(5):98. https://doi.org/10.5688/aj700598

Shank RA, Foroud NA, Hazendonk P, Eudes F, Blackwell BA (2011) Current and future experimental strategies for structural analysis of trichothecene mycotoxins - a prospectus. Toxins 3(12):1518–1553. https://doi.org/10.3390/toxins3121518

Shephard GS, Sewram V, Nieuwoudt TW, Marasas WF, Ritieni A (1999) Production of the mycotoxins fusaproliferin and beauvericin by South African isolates in the *Fusarium* section Liseola. J Agric Food Chem 47(12):5111–5115. https://doi.org/10.1021/jf9903713

Stange K, Pohlmeier H, Lubbesmeyer A, Gumbinger G, Schmitz W, Baumgart P (1998) Ergotamine-induced vascular spasms through chronic inhalation of ergotamine during the preparation of rye flour. Dtsch Med Wochenschr 123:1547–1550

Stoloff L (1976) Occurrence of mycotoxins in foods. In: Selected technical publications, vol 261. National Technical Information Service, Springfield, VA

Stott WT, Bullerman LB (1975) Patulin: a mycotoxin of potential concern in foods. J Milk Food Technol 38(11):695–705. https://doi.org/10.4315/0022-2747-38.11.695

Sudakin DL (2003) Trichothecenes in the environment: relevance to human health. Toxicol Lett 143(2):97–107. https://doi.org/10.1016/S0378-4274(03)00116-4

Sy-Cordero AA, Pearce CJ, Oberlies NH (2012) Revisiting the enniatins: a review of their isolation, biosynthesis, structure determination and biological activities. J Antibiot 65(11):541–549

Tatsuno T, Saito M, Enomoto M, Tsunoda H (1968) Nivalenol, a toxic principle of *Fusarium nivale*. Chem Pharm Bull 16(12):2519–2520. https://doi.org/10.1248/cpb.16.2519

Tyko K (2021) Dog food recall expands: more than 70 dogs have died and 80 pets sick after eating Sportsmix pet food. USA Today. Accessed 13 Jan 2021

Uraguchi K (1967) Mycotoxins of "yellowed rice". Asian Med J 10(4):41–44
Van Dongen PW, de Groot AN (1995) History of ergot alkaloids from ergotism to ergometrine. Eur J Obstet Gynecol Reprod Biol 60(2):109–116. https://doi.org/10.1016/0028-2243(95)02104-Z
Van Luijk A (1938) Antagonism of *Penicillium* spp. versus *Pythium debaryanum*. Chron Bot 4:210
Weber J, Vaclavikova M, Wiesenberger G, Haider M, Hametner C, Fröhlich J, Berthiller F, Adam G, Mikula H, Fruhmann P (2018) Chemical synthesis of culmorin metabolites and their biologic role in culmorin and acetyl-culmorin treated wheat cells. Org Biomol Chem 16(12):2043–2048. https://doi.org/10.1039/C7OB02460F
Weindlin R, Emerson O (1936) Phytopathology 26:1068
Yabuta T, Kambe K, Hayashi T (1934) Biochemical studies of bakanae fungus on rice. I. Fusarinic acid, a new product of the bakanae fungus. J Agric Chem Soc Jpn 10:1059–1068. Fusaric acid. The Merck Index Edition 15, 2013, 793 https://www.bioaustralis.com/product/fusaric-acid/
Yazar S, Omurtag GZ (2008) Fumonisins, trichothecenes and zearalenone in cereals. Int J Mol Sci 9 (11):2062–2090
Zain ME (2011) Impact of mycotoxins on humans and animals. J Saudi Chem Soc 15(2):129–144. https://doi.org/10.1016/j.jscs.2010.06.006

Chapter 2
Overview

2.1 Introduction

Cereal grains and associated by-products constitute important sources of energy and protein for all classes of farm livestock and humans. When moulds colonize cereal grains and animal feed, there is a significant risk of contamination with the secondary metabolites of these fungi, which are known as mycotoxins (Bennett and Klich 2003). Mycotoxins are predominantly produced by the mycelial structures of filamentous fungi, the moulds (Hussein and Brasel 2001). They are a diverse range of low-molecular-weight natural products that are harmful to animals and humans (Akande et al. 2006). Mycotoxins differ from each other in their toxicity as well as chemical structure; nevertheless, they are grouped together because they cause disease and death in human beings as well as in other vertebrates.

Grain is the main constituent of the energy-rich diet consumed by farm animals raised in integrated production systems. Filamentous fungi commonly growing on these grains produce mycotoxins, which is one of the drawbacks of using these grains in feed formulations (Diekman and Green 1992). Approximately 70% of the feed based on cereals is contaminated with a minimum of one mycotoxin (Streit et al. 2013). Mycotoxins thus enter the food chain and evoke powerful toxic responses even at very low concentrations in animals and further in humans (Bennett 1987). Over 300 mycotoxins are known to date, and aflatoxin, fumonisin, ochratoxin, trichothecene (deoxynivalenol, T-2, and HT-2), and ZEN are the most commonly reported mycotoxins due to their safety concerns and economic importance (Murugesan et al. 2015; Pereira et al. 2019). Table 2.1 lists some mycotoxins and the organisms by which they are synthesized.

K. Singh, A. Kumari, *Mycotoxins and Mycotoxicoses*,
https://doi.org/10.1007/978-981-19-2370-8_2

Table 2.1 A partial listing of mycotoxins and their producers

Mycotoxins	Organism	Reference
Aflatoxins and structurally related mycotoxins	*Aspergillus flavus*	Asao et al. (1963)
Zearalenone and zearalenol	*Fusarium epispheria, F. moniliforme, Gibberella fujikuroi, F. nivale, F. oxysporwn, F. roseum, F. solani, F. tricinctum*	Caldwell et al. (1970)
Moniliformin	*F. moniliforme*	Marasas et al. (1986)
Sterigmatocystins	*Aspergillus versicolor, A. chevalieri, A. ruber, A. amstelodami, A. aureolatus, A. quadrilineatus, A. sydowi*	Veršilovskis and De Saeger (2010)
Butenolide	*Aspergillus terreus*	Elkhayat et al. (2016)
Versicolorins	*Aspergillus parasiticus*	Bennett et al. (1976)
Fusaric acid	*F. crookwellense, F. subglutinans, F. sambucinum, F. napiforme, F. heterosporum, F. oxysporum, F. solan*	Bacon et al. (1996)
Aspertoxins	*Aspergillus flavus*	Rodricks et al. (1968)
Fusarin A–D	Fusarin C: *Fusarium moniliforme* and *Fusarium venenatum*	Song et al. (2004)
Beauvericin	*F. culmorum, F. cerealis, F. graminearum, F. sambucinum, F. venenatum, F. torulosum, F. flocciferum, F. acuminatum var. acuminatum, F. compactum, F. scirpi, F. equiseti, F. longipes, F. subglutinans, F. anthophilum, F. chlamydosporum, F. poae, F. sporotrichioides, F. tricinctum, F. polyphialidicum, F. beomiforme, F. dlamini, F. nygamai*	Logrieco et al. (1998)
Sterigmatin	*Aspergillus versicolor*	Fukuyama et al. (1975)
Fusaproliferin	*Fusarium proliferatum*	Ritieni et al. (1995)
Gliotoxin	*Aspergillus fumigates*	Carberry et al. (2012)
Averufarin	*Beauveria bassiana*	Cannell et al. (1997)
Curvularin	*Curvularia* sp.	Dai et al. (2010)
Luteoskyrin	*Penicillium islandicum*	Ueno and Ishikawa (1969)
Alternariol	*Alternaria*	Lou et al. (2016)
Patulin	*Aspergillus, Byssochlamys*, and *Penicillium griseofulvum*	Saleh and Goktepe (2019), Raistrick (1943)
Citrinin	*Penicillium citrinum*	Hetherington and Raistrick (1931)
Penicillic acid	*Pinctada martensii, Penicillium palitans, Penicillium cyclopium, Pen icillium puberulum*	Wilson et al. (1968), Ciegler and Kurtzman (1970)

(continued)

Table 2.1 (continued)

Mycotoxins	Organism	Reference
Citreoviridin	*Eupenicillium ochrosalmoneum*	Wicklow et al. (1988)
Phomopsins	*Phomopsis* sp.	Huang et al. (2008)
Cyclopiazonic acid	*Penicillium cyclopium*, *Aspergillus versicolor*, *A. oryzae*, *A. flavus*, *A. tamarii*, and *Penicillium camembertii*	Holzapfel (1968), Ohmomo et al. (1973), Lorenz et al. (2007), Luk et al. (1977), Dorner (1983), Still et al. (1978)
Cyclosporine	*Neocosmospora vasinfecta*	Nakajima et al. (1989)
Diplonene	*Stenocarpella*	Haschek et al. (2013)
Ergot alkaloids	*Aspergillus fumigatus*, *Claviceps purpurea*, *Clavulinopsis fusiformis*, *Neotyphodium lolli*, *Arthroderma benhamiae*	Coyle and Panaccione (2005), Haarmann et al. (2005), Lorenz et al. (2007), Jakubczyk et al. (2014), Tasker and Wipf (2020)
Tremorgens	*Aspergillus flavus*, *Claviceps paspali*	Wilson and Wilson (1964), Gallagher et al. (1980)
Ergotamine	*Claviceps purpurea*	Haarmann et al. (2008)
Aflatrem	*Aspergillus oryzae*	Tagami et al. (2014)
Penitrem A	*Penicillium crustosum*, *P. palitans*, *P. carneum*, *P. simplicissimum*	Tagami et al. (2014), Hou et al. (1971), Boysen et al. (1996), Hayashi et al. (1993)
Roquefortine	*Penicillium granulatum*	Niu et al. (2018)
Paxilline	*Penicillium paxilli*	Saikia et al. (2006)
Trichothecenes	*Fusarium*, *Myrothecium*, *Spicellum*, *Stachybotrys*, *Cephalosporium*, *Trichoderma*, *Trichothecium*	Xu et al. (2006), McCormick et al. (2011)
Verruculogen	*Penicillium verruculosum*, *Aspergillus fumigatus*	Fayos et al. (1974), Steffan et al. (2009)
Deoxynivalenol (DON)	*Fusarium* spp.	Audenaert et al. (2014)
Fumigaclavine	*Aspergillus fumigates*	Li et al. (2013)
Verrucarins	*Myrothecium verrucaria*	McPhail and Sim (1966)
Slaframine	*Rhizoctonia leguminicola*	Harris et al. (1987)
Roridins	*Myrothecium roridum*	Xu et al. (2006)
Satratoxins	*Stachybotrys chartarum*	McPhail and Sim (1966)
Tenuazonic acid	*Alternaria alternata*	Kumari and Tirkey (2019)
Bipolarin	*Aspergillus versicolor*	Abdel-Wahab et al. (2019)
Ochratoxin A	*A. clavatus*, *A. ochraceus*, *A. versicolor*, and *A. wentii*	Varga et al. (2010)
Atranones A–G	*Stachybotrys chartarum*	Hinkley et al. (2000)
Enniatins	*Fusarium* species	Sy-Cordero et al. (2012)
Pyrrocidines	*Acremonium zeae*	Wicklow and Poling (2009)
Rubratoxins A and B	*Penicillium rubrum*	Moss et al. (1971)
Sporodesmin	*Pithomyces chartarum*	Brewer et al. (1989)
Ergonovine	*Achnatherum inebrians*	Miles et al. (1996)
Fusarium toxins	*Fusarium* species	Lappalainen et al. (1996)

2.2 Mycotoxicoses

Mycotoxicoses are the animal or human diseases caused by ingestion, inhalation, or skin contact with mycotoxin (Kumari 2018). There are various effects of mycotoxins in animals and humans, ranging from suppression of the immune system to being fatal in rigorous conditions. The severity of illness is toxin-related (type of mycotoxin consumed, level and period of time for which the mycotoxin is taken), animal-related (animal age, breed, general health, immune status, species, sex, nutritional standing), and environment-related (temperature, farm management, hygiene) (Heidler and Schatzmayr 2003). The negative effects of mycotoxins are magnified in poor management and health conditions on a farm. When pigs are considered, piglets and female breeders are the most vulnerable animals, but attention should be paid to final animal products as mycotoxins can be carried over to blood and muscles and appear in the meat. To avoid this and to reduce the incidence of mycotoxins in human food, limits regarding the permissible concentration of mycotoxins have been imposed worldwide in cereal grains, animal feeds, etc. (Biomin n.d., http://www.biomin.net/en/knowledge-center/).

Mycotoxins cause a wide array of illnesses in humans ranging from allergy to immuno-suppression and cancer. Aflatoxins, ochratoxin A, fumonisins, trichothecenes, and zearalenone belong to the list of most significant mycotoxins. The first and most important of all are the aflatoxins. They are hepatotoxic and carcinogenic and, in combination with hepatitis B virus, result in several thousand deaths every year mostly in tropical countries and non-industrialized nations. Ochratoxin A is another mycotoxin having carcinogenic properties. It is associated with urinary tract cancer and kidney damage in northern and Eastern Europe. Oesophageal cancer caused by fumonisins has been reported in southern Africa, China, and some other places. Trichothecenes show immunosuppression, while zearalenone shows oestrogenic effects in humans and animals (Pitt 2000). Table 2.2 enlists some major mycotoxins and the organ system affected by them.

Table 2.2 Some major mycotoxins and the organ system affected by them

Toxin(s)	Organ system affected
Aflatoxin	Vascular
Aflatoxin, T-2 toxin, DON	Digestive
Trichothecenes	Respiratory
Trichothecenes	Nervous
Trichothecenes	Cutaneous
Ochratoxin A, citrinin	Urinary
Zearalenone, T-2 toxin	Reproductive
Aflatoxin, ochratoxin, etc.	Immune

2.3 Mycotoxins in Indoor Environment

The growth of fungi in a damp indoor environment has been gathering importance over the past decade. Researchers have reported that children and adults exposed to fungi growing in damp indoors are affected by respiratory and other problems like headache, nausea, tiredness, and vomiting (Richard 2007). *Aspergillus versicolor* is a mould that is commonly present in indoor air that causes inflammation and toxicity in lungs (Jussila et al. 2002). Studies reveal that aflatoxins released by *Aspergillus flavus* turn out to be carcinogenic and immunosuppressive in animals when inhaled (Zarba et al. 1992; Louria et al. 1974). Moreover, *Stachybotrys* in an indoor environment can cause haemorrhage and pulmonary haemosiderosis in infants and many other health problems in adults (Croft et al. 1986; Hodgson et al. 1998; Engelhart et al. 2002).

2.4 Mushroom Toxins

Although most mushrooms are harmless, some of them are poisonous and contain mycotoxins capable of attacking the human host with fatal consequences. The reaction time and symptoms of mycotoxicoses may vary. Some mycotoxins act immediately on the human host, with effects ranging from anxiety, diarrhoea, hallucination, muscle spasms, nausea, vomiting to hyperactivity or lethargy. The sudden symptoms are caused by the mycotoxins coprine, ibotenic acid, muscarine, and psilocybin/psilocin. On the other hand, some mycotoxins have more serious effects, are delayed in commencement, and can be lethal. The late symptoms (which may continue up to 14 days after ingestion) include bloating, cramps, diarrhoea, frequent urination, headache, kidney pain, severe vomiting, severe thirst, and death. Monomethylhydrazine, amatoxins, and orellanine are potentially fatal mycotoxins that are classified as group A, B, and C poisons (Science Encyclopedia n.d., http://science.jrank.org/pages/4539/Mycotoxin.html). Table 2.3 lists mushroom mycotoxins, the organism that produces them, and their health effects.

2.5 Biomarkers

Mycotoxins possess a wide range of chemical structures and mechanisms of action. Any organ system can be the biological target for mycotoxins. Qualitative and quantitative assessment of exposure, internal dose, toxicological effects, and susceptibility can be facilitated by biomarkers.

Mycotoxin exposure can be successfully documented by the assay of mycotoxins in the fluids of the body and tissues. The concentration of the mycotoxins and their metabolites present inside the body is termed an internal dose. Apart from exposure

Table 2.3 Toxins produced by mushrooms and their health effects (Feng et al. 2015; Yin et al. 2014)

Organism	Toxin produced	Health effects
Amanita phalloides	Amatoxins, phallotoxins, virotoxins	Causes extreme gastrointestinal upset
Tricholoma terreum	Fifteen triterpenoid terreolides and saponaceolides	Cytotoxicity in cancer cell lines, rhabdomyolysis
Gyromitra esculenta, *G. gigas*, *G. fastigiata*	Gyromitrin hydrolysed in body monohydrazinemethyl	Causes brain damage, seizures, gastrointestinal upset, and haemolysis; metabolic poison; the principal toxin in the genus *Gyromitra*; antidote is large doses of intravenous pyridoxine hydrochloride
Cortinarius	Orellanine	Redox cycler similar to paraquat; causes kidney failure within 3 weeks after ingestion
Inocybe and *Clitocybe* species such as *C. dealbata*	Muscarine	Causes SLUDGE syndrome. Found in various mushrooms; antidote is atropine
Psilocybe azurescens, *P. semilanceata*, *P. cyanescens*	Psilocybin	Causes CNS arousal and hallucinations
Amanita muscaria and related species	Muscimol and ibotenic acid	Excitotoxin
Coprinopsis atramentaria	Coprine	Causes illness when consumed with alcohol

and toxicity studies related to mycotoxins, toxicokinetic studies are also very crucial for accurate risk assessment as oral bioavailability, absorption rate, pre-systemic biotransformation, distribution, and elimination. These are important factors that affect the internal dose of a toxicant (Diekman and Green 1992; Hussein and Brasel 2001).

2.6 Mycotoxins Worldwide

A large portion of the world's food like cereals, groundnuts, maize, and tree nuts are known to be contaminated with aflatoxins and fumonisins (FB). Co-occurrence of these two mycotoxins is often come across in maize. In places like Asia, Africa, and Latin America where these commodities form a staple diet, their contamination leads to elevated chronic exposures. It is most commonly found in populations that are based on farming as regulations and control policies here are either not present or are not enforced. Aflatoxins are hepatocarcinogenic in humans, and in combination with chronic hepatitis B virus infection, it results in aflatoxicosis outbreaks. Aflatoxin consumption leads to impaired growth in animals and immuno-suppression in humans and animals. Immuno-suppression has gained importance amongst the population as it makes an individual susceptible to other diseases and infections. Fumonisins B (FB) has been grouped as Group 2B (possibly carcinogenic to humans) because studies on rodents have shown that they induce the formation of tumours in the liver and kidney. It has also been associated with increasing oesophageal cancer in previous outbreaks. Studies on FB also suggest that it interferes with the biosynthesis of sphingolipid and ceramide synthase leading to neural tube defects. These defects have been recorded in populations consuming maize contaminated with FB. The data available through the studies carried out worldwide is enough to put a regulation on the limit of exposure amongst susceptible populations (Wild and Gong 2010).

Aflatoxins were first identified in the 1960s, and since it was discovered, many nations have implied regulations to protect their consumers from the ill effects of aflatoxin contamination and to make certain that the food being exported to other countries is within the permissible limits (Reddy and Raghavender 2007). The important features that are considered while laying down regulatory limits include the availability of scientific data related to toxicity, information regarding the distribution and level of consumption of commodities, and methods of detection and analysis. Another important factor taken into consideration is the economic factors that include food security issues and commercial and trade interests of the nation. Considering all the factors, mycotoxin tolerance limits are set up (Reddy and Raghavender 2007).

In spite of being a complicated issue, many nations have set up permissible limits in the past decades, and many new regulations are still coming into play. Aflatoxins and aflatoxin M1, deoxynivalenol, diacetoxyscirpenol, fumonisins B1, B2, and B3, ochratoxin A, trichothecenes T-2 toxin and HT-2 toxin, zearalenone, ergot alkaloids, patulin, sterigmatocystin, and phomopsins are some natural mycotoxin for which permissible limits have been established by several countries (El Khoury and Atoui 2010).

There are several published data provided by the authorities responsible for the implementation of the rules on the legalization of regulatory limits, permissible limits, protocols on sampling, and analysis of mycotoxins (Reddy and Raghavender 2007). Various international bodies ascertain the fact that food and feed being consumed and exported follow the existing legislation. International bodies like the Food and Agriculture Organization of the United Nations (FAO), the European Union (EU), the Codex Alimentarius Commission (CAC), and the World Health Organization (WHO) keep providing data on the worldwide legislation of mycotoxins. The number of countries having set up permissible limits for mycotoxins keeps increases every year, which indicates that governments of these countries are now becoming alert about the harmful effects of mycotoxin contamination and its repercussions on human and animal health and international trade.

A study carried out for the United States, using computer simulations and considering the data provided by the FDA, showed that every year crop worth of $932 million is lost to mycotoxin contamination when only aflatoxin, deoxynivalenol, and fumonisin were considered in the contamination of wheat, corn, and peanuts (Dohlman 2003). Other factors like quality control, testing, and regulations to limit or prevent contamination of mycotoxins lead to an additional cost of $466 million. Other reports also stated that the losses incurred by the poultry and swine industry in the United States in some years had exceeded $100 million (Dohlman 2003).

2.7 Conclusion

Exposure of mycotoxins is a serious issue worldwide. Every known plant and animal product is susceptible to attack by moulds under appropriate conditions, but the number of known mycotoxins of practical importance is relatively small. It is now known for decades that mycotoxin contamination causes harmful effects, but we have still not been able to eradicate even the oldest known mycotoxicosis, ergotism, and are still on the way to controlling mycotoxins. The occurrence of mycotoxicosis is more widespread than it is reported. Since perfect safety may be an unattainable dream, we may have to learn to live with an acceptable level of risk.

References

Abdel-Wahab NM, Scharf S, Özkaya FC, Kurtán T, Mándi A, Fouad MA, Kamel MS, Müller WE, Kalscheuer R, Lin W, Daletos G (2019) Induction of secondary metabolites from the marine-derived fungus *Aspergillus versicolor* through co-cultivation with *Bacillus subtilis*. Planta Med 85(06):503–512

Akande KE, Abubakar MM, Adegbola TA, Bogoro SE (2006) Nutritional and health implications of mycotoxins in animal feeds: a review. Pak J Nutr 5(5):398–403

Asao T, Buchi G, Abdel-Kader MM, Chang SB, Wick EL, Wogan GN (1963) Aflatoxins B and G. J Am Chem Soc 85(11):1706–1707. https://doi.org/10.1021/ja00894a050

Audenaert K, Vanheule A, Höfte M, Haesaert G (2014) Deoxynivalenol: a major player in the multifaceted response of *Fusarium* to its environment. Toxins 6(1):1–9. https://doi.org/10.3390/toxins6010001

Bacon CW, Porter JK, Norred WP, Leslie J (1996) Production of fusaric acid by *Fusarium* species. Appl Environ Microbiol 62(11):4039–4043. https://doi.org/10.1128/aem.62.11.4039-4043.1996

Bennett JW (1987) Mycotoxins, mycotoxicoses, mycotoxicology and mycopathologia. Mycopathologia 100(1):3–5

Bennett JW, Klich M (2003) Mycotoxins. Clin Microbiol Rev 16:497–516. https://doi.org/10.1128/CMR.16.3.497-516.2003

Bennett JW, Lee LS, Cucullu AF (1976) Effect of dichlorvos on aflatoxin and versicolorin A production in *Aspergillus parasiticus*. Bot Gaz 137(4):318–324

Biomin (n.d.) Mycotoxins, Knowledge Centre. http://www.biomin.net/en/knowledge-center/. Accessed 25 Sep 2021

Boysen M, Skouboe P, Frisvad J, Rossen L (1996) Reclassification of the *Penicillium roqueforti* group into three species on the basis of molecular genetic and biochemical profiles. Microbiology 142(3):541–549. https://doi.org/10.1099/13500872-142-3-541

Brewer D, Russell DW, de Melio Amaral RE, Aycardi ER (1989) An examination of North and South American isolates of *Pithomyces chartarum* for production of sporidesmin and sporidesmolides. Proc N S Inst Sci 38:73

Caldwell RW, Tuite J, Stob M, Baldwin R (1970) Zearalenone production by *Fusarium* species. Appl Microbiol 20(1):31–34. https://doi.org/10.1128/am.20.1.31-34.1970

Cannell RJ, Rashid T, Ismail IM, Sidebottom PJ, Knaggs AR, Marshall PS (1997) Novel metabolites of warfarin produced by *Beauveria bassiana* and *Streptomyces rimosus*: a novel application of hplc-nmr. Xenobiotica 27(2):147–157. https://doi.org/10.1080/004982597240659

Carberry S, Molloy E, Hammel S, O'Keeffe G, Jones GW, Kavanagh K, Doyle S (2012) Gliotoxin effects on fungal growth: mechanisms and exploitation. Fungal Genet Biol 49(4):302–312. https://doi.org/10.1016/j.fgb.2012.02.003

Ciegler A, Kurtzman CP (1970) Penicillic acid production by blue-eye fungi on various agricultural commodities. Appl Microbiol 20(5):761–764

Coyle CM, Panaccione DG (2005) Appl Environ Microbiol 71:3112–3118

Croft WA, Jarvis BB, Yatawara CS (1986) Airborne outbreak of trichothecenetoxicosis. Atmos Environ 20(3):549–552

Dai J, Krohn K, Flörke U, Pescitelli G, Kerti G, Papp T, Kövér KE, Bényei AC, Draeger S, Schulz B, Kurtán T (2010) Curvularin-type metabolites from the fungus *Curvularia* sp. isolated from a marine alga. Eur J Org Chem 2010:6928–6937. https://doi.org/10.1002/ejoc.201001237

Diekman MA, Green ML (1992) Mycotoxins and reproduction in domestic livestock. J Anim Sci 70(5):1615–1627

Dohlman E (2003) Mycotoxin hazards and regulations. Int Trade Food Saf 828:97

Dorner JW (1983) Production of cyclopiazonic acid by *Aspergillus tamarii* Kita. Appl Environ Microbiol 46(6):1435–1437. https://doi.org/10.1128/aem.46.6.1435-1437.1983

El Khoury A, Atoui A (2010) Ochratoxin A: general overview and actual molecular status. Toxins 2(4):461–493

Elkhayat ES, Ibrahim SR, Mohamed GA, Ross SA (2016) Terrenolide S, a new antileishmanial butenolide from the endophytic fungus *Aspergillus terreus*. Nat Prod Res 30(7):814–820. https://doi.org/10.1080/14786419.2015.1072711

Engelhart S, Loock A, Skutlarek D, Sagunski H, Lommel A, Färber H, Exner M (2002) Occurrence of toxigenic Aspergillusversicolor isolates and sterigmatocystin in carpet dust from damp indoor environments. Appl Environ Microbiol 68(8):3886–3890

Fayos J, Lokensgard D, Clardy J, Cole RJ, Kirksey JW (1974) Structure of verruculogen, a tremor producing peroxide from *Penicillium verruculosum*. J Am Chem Soc 96(21):6785–6787. https://doi.org/10.1021/ja00828a054

Feng T, He J, Ai HL, Huang R, Li ZH, Liu JK (2015) Three New Triterpenoids from European Mushroom *Tricholoma terreum*. Nat Prod Bioprospect 5(4):205–208

Fukuyama K, Tsukihara T, Katsube Y, Hamasaki T, Hatsuda Y, Tanaka N, Ashida T, Kakudo M (1975) The crystal and molecular structure of sterigmatin, a metabolite of *Aspergillus versicolor*. Bull Chem Soc Jpn 48(5):1639–1640. https://doi.org/10.1246/bcsj.48.1639

Gallagher RT, Clardy J, Wilson BJ (1980) Aflatrem, a tremorgenic toxin from *Aspergillus flavus*. Tetrahedron Lett 21(3):239–242. https://doi.org/10.1016/S0040-4039(00)71178-6

Haarmann T, Machado C, Lübbe Y, Correia T, Schardl CL, Panaccione DG, Tudzynski P (2005) The ergot alkaloid gene cluster in *Claviceps purpurea*: extension of the cluster sequence and intra species evolution. Phytochemistry 66(11):1312–1320. https://doi.org/10.1016/j.phytochem.2005.04.011Get

Haarmann T, Lorenz N, Tudzynski P (2008) Use of a nonhomologous end joining deficient strain (Δku70) of the ergot fungus *Claviceps purpurea* for identification of a nonribosomal peptide synthetase gene involved in ergotamine biosynthesis. Fungal Genet Biol 45(1):35–44. https://doi.org/10.1016/j.fgb.2007.04.008

Harris TM, Harris CM, Hill JE, Ungemach FS, Broquist HP, Wickwire BM (1987) (1S, 2R, 8aS)-1, 2-Dihydroxyindolizidine formation by *Rhizoctonia leguminicola*, the fungus that produces slaframine and swainsonine. J Org Chem 52(14):3094–3098. https://doi.org/10.1021/jo00390a024

Haschek WM, Rousseaux CG, Wallig MA, Bolon B, Ochoa R (eds) (2013) Haschek and Rousseaux's handbook of toxicologic pathology. Academic Press, London

Hayashi H, Asabu Y, Murao S, Nakayama M, Arai M (1993) A new congener of penitrems, 6-bromopenitrem E, from *Penicillium simplicissimum* AK-40. ChemInform 24(31)

Heidler D, Schatzmayr G (2003) A new approach to managing mycotoxins. Feed Mix 11(1):31–34

Hetherington AC, Raistrick H (1931) On the production and chemical constitution of a new yellow colouring matter, citrinin, produced from glucose by *Penicillium citrinum* Thom. Phil Trans R Soc Lond B Biol Sci 220:269–295

Hinkley SF, Mazzola EP, Fettinger JC, Lam YF, Jarvis BB (2000) Atranones A–G, from the toxigenic mold *Stachybotrys chartarum*. Phytochemistry 55(6):663–673. https://doi.org/10.1016/S0031-9422(00)00231-4

Hodgson MJ, Morey P, Leung WY, Morrow L, Miller D, Jarvis BB, Storey E (1998) Building-Associated Pulmonary Disease from Exposure to *Stachybotrys chartarum* and *Aspergillus versicolor*. J Occup Environ Med 40(3):241–249

Holzapfel CW (1968) The isolation and structure of cyclopiazonic acid, a toxic metabolite of *Penicillium cyclopium* Westling. Tetrahedron 24(5):2101–2119. https://doi.org/10.1016/0040-4020(68)88113-X

Hou CT, Ciegler A, Hesseltine CW (1971) Tremorgenic toxins from Penicillia: III. tremortin production by *Penicillium* species on various agricultural commodities. Appl Microbiol 21(6):1101–1103. https://doi.org/10.1128/am.21.6.1101-1103.1971

Huang Z, Cai X, Shao C, She Z, Xia X, Chen Y, Yang J, Zhou S, Lin Y (2008) Chemistry and weak antimicrobial activities of phomopsins produced by mangrove endophytic fungus *Phomopsis* sp. ZSU-H76. Phytochemistry 69(7):1604–1608. https://doi.org/10.1016/j.phytochem.2008.02.002

Hussein HS, Brasel JM (2001) Toxicity, metabolism, and impact of mycotoxins on humans and animals. Toxicology 167(2):101–134. https://doi.org/10.1016/S0300-483X(01)00471-1

Jakubczyk D, Cheng JZ, O'Connor SE (2014) Biosynthesis of the ergot alkaloids. Nat Prod Rep 31(10):1328–1338

Jussila J, Komulainen H, Kosma VM, Nevalainen A, Pelkonen J, Hirvonen MR (2002) Spores of Aspergillusversicolor isolated from indoor air of a moisture-damaged building provoke acute inflammation in mouse lungs. Inhal Toxicol 14(12):1261–1277

Kumari A (2018) Mycotoxicoses: a repercussion of unhygienic environmental conditions. In: Pathak AM, Singh K (eds) Bhartiya sanskriti mei prakriti: Ek anusheelan. Kala Prakashan, Delhi, pp 132–141. ISBN 9789387199033

Kumari A, Tirkey NN (2019) Tenuazonic acid: a potent mycotoxin. In: Singh K, Srivastav AN (eds) Recent trends in human and animal mycology. Springer, Singapore, pp 203–211

Lappalainen S, Nikulin M, Berg S, Parikka P, Hintikka EL, Pasanen AL (1996) Fusarium toxins and fungi associated with handling of grain on eight Finnish farms. Atmos Environ 30(17):3059–3065

Li YX, Himaya SW, Dewapriya P, Zhang C, Kim SK (2013) Fumigaclavine C from a marine-derived fungus *Aspergillus fumigatus i*nduces apoptosis in MCF-7 breast cancer cells. Mar Drugs 11(12):5063–5086. https://doi.org/10.3390/md11125063

Logrieco A, Moretti A, Castella G, Kostecki M, Golinski P, Ritieni A, Chelkowski J (1998) Beauvericin production by *Fusarium* species. Appl Environ Microbiol 64(8):3084–3088. https://doi.org/10.1128/AEM.64.8.3084-3088.1998

Lorenz N, Wilson EV, Machado C, Schardl CL, Tudzynski P (2007) Appl Environ Microbiol 73:7185–7191

Lou J, Yu R, Wang X, Mao Z, Fu L, Liu Y, Zhou L (2016) Alternariol 9-methyl ether from the endophytic fungus *Alternaria* sp. Samif01 and its bioactivities. Braz J Microbiol 47:96–101. https://doi.org/10.1016/j.bjm.2015.11.004

Louria DB, Finkel G, Smith JK, Buse M (1974) Aflatoxin-induced tumors in mice. Sabouraudia 12(3):371–375

Luk KC, Kobbe B, Townsend JM (1977) Production of cyclopiazonic acid by *Aspergillus flavus* Link. Appl Environ Microbiol 33(1):211–212. https://doi.org/10.1128/aem.33.1.211-212.1977

Marasas WF, Thiel PG, Rabie CJ, Nelson PE, Toussoun TA (1986) Moniliformin production in *Fusarium* section Liseola. Mycologia 78(2):242–247. https://doi.org/10.1080/00275514.1986.12025235

McCormick SP, Stanley AM, Stover NA, Alexander NJ (2011) Trichothecenes: from simple to complex mycotoxins. Toxins 3(7):802–814. https://doi.org/10.3390/toxins3070802

McPhail AT, Sim GA (1966) Fungal metabolites. Part VI. The structure of verrucarin A: X-ray analysis of verrucarin A p-iodobenzenesulphonate. J Chem Soc C Org:1394–1406

Miles CO, Lane GA, di Menna ME, Garthwaite I, Piper EL, Ball OJ, Latch GC, Allen JM, Hunt MB, Bush LP, Min FK (1996) High levels of ergonovine and lysergic acid amide in toxic Achnatherum inebrians accompany infection by an *Acremonium*-like endophytic fungus. J Agric Food Chem 44(5):1285–1290

Moss MO, Robinson FV, Wood AB (1971) Rubratoxins. J Chem Soc C Org:619–624

Murugesan GR, Ledoux DR, Naehrer K, Berthiller F, Applegate TJ, Grenier B, Phillips TD, Schatzmayr G (2015) Prevalence and effects of mycotoxins on poultry health and performance, and recent development in mycotoxin counteracting strategies. Poult Sci 94(6):1298–1315

Nakajima H, Hamasaki T, Tanaka K, Kimura Y, Udagawa SI, Horie Y (1989) Production of cyclosporin by fungi belonging to the genus *Neocosmospora*. Agric Biol Chem 53(8): 2291–2292. https://doi.org/10.1080/00021369.1989.10869637

Niu S, Wang N, Xie CL, Fan Z, Luo Z, Chen HF, Yang XW (2018) Roquefortine J, a novel roquefortine alkaloid, from the deep-sea-derived fungus *Penicillium granulatum* MCCC 3A00475. J Antibiot 71(7):658–661

Ohmomo SM, Sugita M, Abe M (1973) Isolation of cyclopiazonic acid, cyclopiazonic acid imine and hissecodehydro cyclopiazonic acid from the cultures of *Aspergillus versicolor* (Vull) Tiraboschi. J Agric Chem Soc Jpn 47:83–89

Pereira CS, Cunha SC, Fernandes JO (2019) Prevalent mycotoxins in animal feed: occurrence and analytical methods. Toxins 11(5):290

Pitt JI (2000) Toxigenic fungi and mycotoxins. Br Med Bull 56(1):184–192

Raistrick H (1943) Patulin in the common cold collaborative research on a derivative of *Penicillium patulum Bainier*: I. Introduction. Lancet 242(6273):625. https://doi.org/10.1016/S0140-6736(00)88176-3

Reddy BN, Raghavender CR (2007) Outbreaks of aflatoxicoses in India. Afr J Food Agric Nutr Dev 7(5):1

Richard JL (2007) Some major mycotoxins and their mycotoxicoses-An overview. Int J Food Microbiol 119(1):3–10

Ritieni A, Fogliano V, Randazzo G, Scarallo A, Logrieco A, Moretti A, Manndina L, Bottalico A (1995) Isolation and characterization of fusaproliferin, a new toxic metabolite from *Fusarium proliferatum*. Nat Toxins 3(1):1720. https://doi.org/10.1002/nt.2620030105

Rodricks JV, Henery-Logan KR, Campbell AD, Stoloff L, Verrett MJ (1968) Isolation of a New Toxin from Cultures of *Aspergillus flavus*. Nature 217(5129):668–668

Saikia S, Parker EJ, Koulman A, Scott B (2006) Four gene products are required for the fungal synthesis of the indole-diterpene, paspaline. FEBS Lett 580(6):1625–1630

Saleh I, Goktepe I (2019) The characteristics, occurrence, and toxicological effects of patulin. Food Chem Toxicol 129:301–311. https://doi.org/10.1016/j.fct.2019.04.036

Science Encyclopedia (n.d.) Mycotoxins. http://science.jrank.org/pages/4539/Mycotoxin.html

Song Z, Cox RJ, Lazarus CM, Simpson TJ (2004) Fusarin C biosynthesis in *Fusarium moniliforme* and *Fusarium venenatum*. ChemBioChem 5(9):1196–1203. https://doi.org/10.1002/cbic.200400138

Steffan N, Grundmann A, Afiyatullov S, Ruan H, Li SM (2009) FtmOx1, a non-heme Fe (II) and α-ketoglutarate-dependent dioxygenase, catalyses the endoperoxide formation of verruculogen in *Aspergillus fumigatus*. Org Biomol Chem 7(19):4082–4087

Still PE, Eckardt C, Leistner L (1978) Bildung yon Cyclopiazonsaure durch *Penicillium camemberti*- Isolate von Kase. Fleischwirtschaft 58:876–877

Streit E, Schwab C, Sulyok M, Naehrer K, Krska R, Schatzmayr G (2013) Multi-mycotoxin screening reveals the occurrence of 139 different secondary metabolites in feed and feed ingredients. Toxins 5(3):504–523

Sy-Cordero AA, Pearce CJ, Oberlies NH (2012) Revisiting the enniatins: a review of their isolation, biosynthesis, structure determination and biological activities. J Antibiot 65(11):541–549

Tagami K, Minami A, Fujii R, Liu C, Tanaka M, Gomi K, Dairi T, Oikawa H (2014) Rapid reconstitution of biosynthetic machinery for fungal metabolites in *Aspergillus oryzae*: total biosynthesis of aflatrem. ChemBioChem 15(14):2076–2080. https://doi.org/10.1002/cbic.201402195

Tasker NR, Wipf P (2020) Biosynthesis, total synthesis, and biological profiles of Ergot alkaloids. Alkaloids Chem Biol 85:1–112. https://doi.org/10.1016/bs.alkal.2020.08.001

Ueno Y, Ishikawa I (1969) Production of luteoskyrin, a hepatotoxic pigment, by *Penicillium islandicum* Sopp. Appl Microbiol 18(3):406–409. https://doi.org/10.1128/am.18.3.406-409.1969

Varga J, Kocsubé S, Péteri Z, Vágvölgyi C, Tóth B (2010) Chemical, physical and biological approaches to prevent ochratoxin induced toxicoses in humans and animals. Toxins 2(7):1718–1750. https://doi.org/10.3390/toxins2071718

Veršilovskis A, De Saeger S (2010) Sterigmatocystin: occurrence in foodstuffs and analytical methods–an overview. Mol Nutr Food Res 54(1):136–147. https://doi.org/10.1002/mnfr.200900345

Wicklow DT, Poling SM (2009) Antimicrobial activity of pyrrocidines from Acremonium zeae against endophytes and pathogens of maize. Phytopathology 99(1):109–115. https://doi.org/10.1094/PHYTO-99-1-0109

Wicklow DT, Stubblefield RD, Horn BW, Shotwell OL (1988) Citreoviridin levels in *Eupenicillium ochrosalmoneum*-infested maize kernels at harvest. Appl Environ Microbiol 54(5):1096–1098. https://doi.org/10.1128/aem.54.5.1096-1098.1988

Wild CP, Gong YY (2010) Mycotoxins and human disease: a largely ignored global health issue. Carcinogenesis 31(1):71–82

Wilson BJ, Wilson CH (1964) Toxin from *Aspergillus flavus*: production on food materials of a substance causing tremors in mice. Science 144(3615):177–178. https://doi.org/10.1126/science.144.3615.177

Wilson BJ, Wilson CH, Hayes AW (1968) Tremorgenic toxin from *Penicillium cyclopium* grown on food materials. Nature 220(5162):77–78

Xu J, Takasaki A, Kobayashi H, Oda T, Yamada J, Mangindaan RE, Ukai K, Nagai H, Namikoshi M (2006) Four new macrocyclic trichothecenes from two strains of marine-derived fungi of the genus Myrothecium. J Antibiot 59(8):451–455

Yin X, Feng T, Shang JH, Zhao YL, Wang F, Li ZH, Dong ZJ, Luo XD, Liu JK (2014) European mushroom Tricholoma terreum. Chem Eur J 20(23):7001–7009

Zarba A, Himieleski R, Hemenway DR, Jakab GJ, Groopman JD (1992) Aflatoxin B1–DNA adduct formation in rat liver following exposure by aerosol inhalation. Carcinogenesis 13(6):1031–1033

Chapter 3
Traditional Mycotoxins and Their Health Implications

3.1 Introduction

The diverse chemical structures and biosynthetic origin of mycotoxins, their numerous biological effects, and production by a large number of fungi make mycotoxins difficult to classify. Clinicians classify them as immunotoxins, nephrotoxins, hepatotoxins, neurotoxins, and so on, according to the organs they affect (Bennett and Klich 2003). Chemists classify them as lactones and coumarins, biochemists as amino acid-derived, polyketides, etc., general practitioners by the illness they cause, and mycologists classify them by the fungi that synthesize them. For example, aflatoxin produced by *Aspergillus flavus*, ochratoxin by *Aspergillus ochraceous*, etc. Traditionally, they are grouped as "field" (or plant pathogenic) and "storage" (or saprophytic) fungi. Some agro-economically important mycotoxins are aflatoxins, ergot alkaloids, fumonisins, tremorgenic toxins, trichothecenes, and zearalenone that contaminate the food and feed during the manufacturing process (Bennet and Klich 2003; Zain 2011).

The occurrence and production of mycotoxins differ due to geographic and climatic diversities, but the exposure to these compounds is universal (Kuiper-Goodman et al. 2010). However, production of mycotoxins for a given species is greatly dependent on growth conditions like humidity, availability of nutrients, and temperature. For example, production of aflatoxins, a mycotoxin produced by *Aspergillus flavus*, occurs over the temperature range of 15–37 °C, but it is hard to indicate the optimum temperature for mycotoxin production (Samapundo et al. 2005).

K. Singh, A. Kumari, *Mycotoxins and Mycotoxicoses*,
https://doi.org/10.1007/978-981-19-2370-8_3

3.2 Aflatoxin

Aflatoxins are the best known and most thoroughly researched mycotoxins worldwide. They are globally associated with a number of diseases known as aflatoxicosis in humans, domestic animals, and livestock. Several environmental factors such as agronomic and agricultural practices, geographic location, and the vulnerability of commodities to fungal invasion and infestation during pre-harvest, storage, and/or processing times influence the presence of aflatoxins and the extent of contamination. The demonstration of the potent carcinogenic effect of aflatoxins and their acute toxicological effects reported in humans and in susceptible laboratory animals has led to greater attention being paid to aflatoxins compared to other mycotoxins (http://www.food-info.net/uk/tox/afla.htm).

3.2.1 Sources

Aflatoxins commonly contaminate food products such as cereal (maize, sorghum, pearl millet, wheat, and rice), oilseeds (cotton, groundnut, sunflower, and soybean), spices (black pepper, chillies, coriander, ginger, and turmeric), tree nuts (almonds, coconut, pistachio, and walnuts), and animal products (milk, cheese, and meat). Cotton seeds, peanuts, and corn are the commodities having the highest risk of aflatoxin contamination (http://www.food-info.net/uk/tox/afla.htm; Reddy and Waliyar 2000; Bbosa et al. 2013).

They are the secondary metabolites of *Aspergillus flavus* and *A. parasiticus*. Other *Aspergillus* species that produce aflatoxins are *A. flavus*, *A.ochraceoroseus*, *A. arachidicola*, *A. minisclerotigenes*, *A. bombycis*, *A. nomius*, *A. rambellii*, *A. parasiticus*, *A. pseudotamarii*, and *Emericella venezuelensis* (Wild and Gong 2009; Park 1993).

3.2.2 Environmental Conditions Favouring the Production of Aflatoxins

A temperature range of 25–30 °C and grain moisture favour the production of aflatoxins. High humidity, drought, stress, and insect damage are associated with their production (Akande et al. 2006).

3.2.3 Physical and Chemical Characteristics

Aflatoxin B1 (Af B1) is the most abundantly found aflatoxin (Fig. 3.1). Af B_1 is a pale-white to yellow crystalline, odourless solid in its pure form. Methanol, chloroform, acetone, and acetonitrile serve as the solvents for aflatoxins (Bbosa et al. 2013). They are dihydrofuran or tetrahydrofuran moieties fused to a coumarin ring. They are fluorescent compounds and are produced via the polyketide pathway in fungi (Akande et al. 2006).

Based on their capacity to fluoresce under UV light (blue or green) and relative chromatographic mobility (subscript) during thin-layer chromatography, aflatoxins are classified as B_1 and B_2 (Figs. 3.1 and 3.2) (producing blue fluorescence) and G_1 and G_2 (producing green fluorescence) (Bennett and Klich 2003) (Figs. 3.3 and 3.4). Two other metabolic products of aflatoxin, M_1 and M_2, are produced by lactating mammals who have consumed aflatoxin-contaminated feed (Figs. 3.5 and 3.6). Aflatoxin B_1 is the most toxic and prevalent amongst all (Biomin 2021a, http://www.mycotoxins.info/en/mycotoxins/).

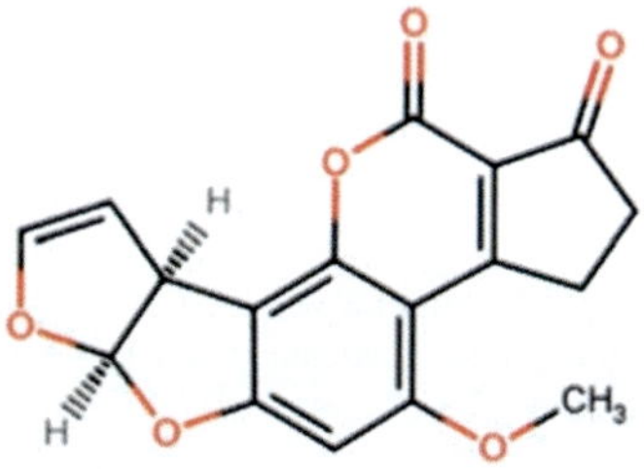

Fig. 3.1 Structure of aflatoxin B_1. (Adapted from Fermentek)

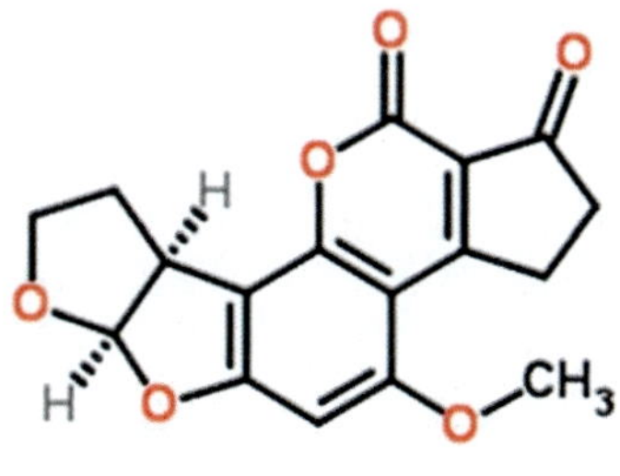

Fig. 3.2 Structure of aflatoxin B_2. (Adapted from Fermentek)

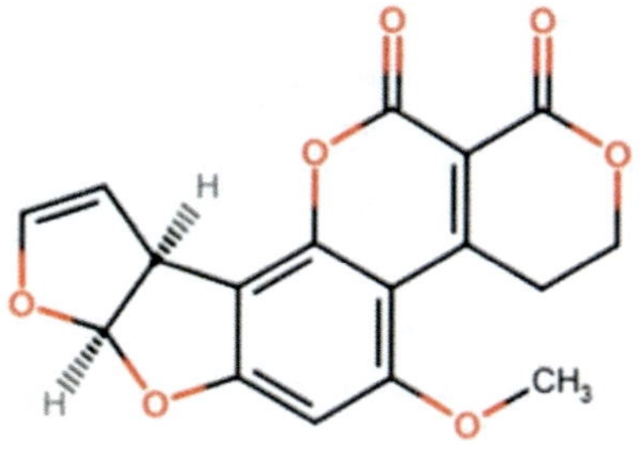

Fig. 3.3 Structure of aflatoxin G_1. (Adapted from Fermentek)

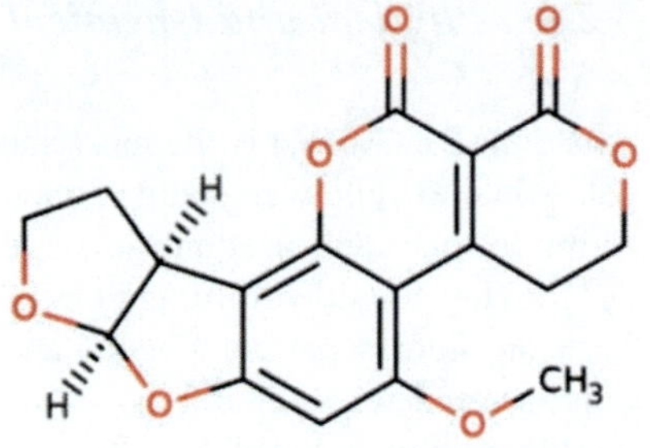

Fig. 3.4 Structure of aflatoxin G_2. (Adapted from Fermentek)

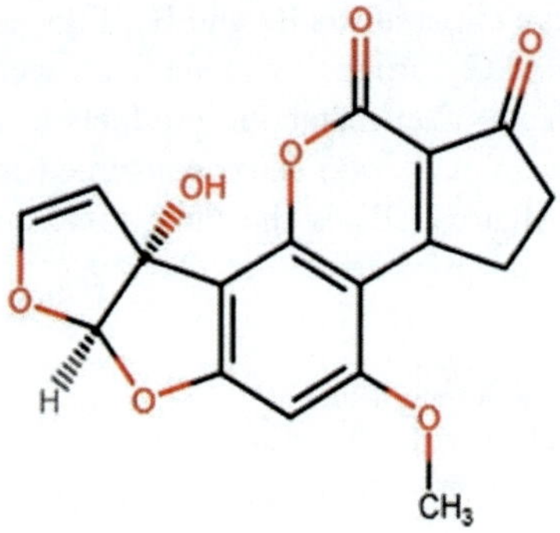

Fig. 3.5 Structure of aflatoxin M_1. (Adapted from Fermentek)

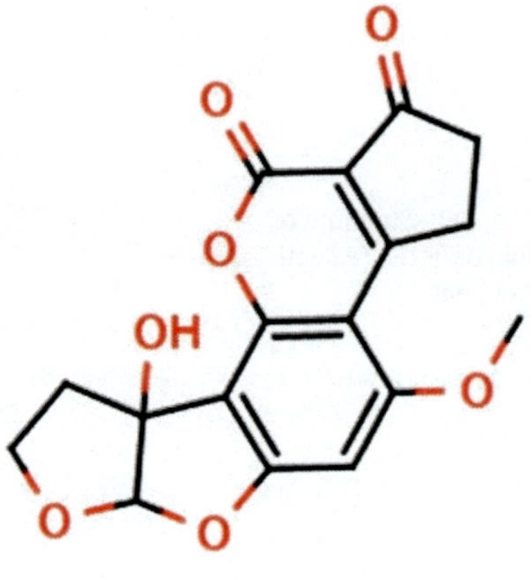

Fig. 3.6 Structure of aflatoxin M_2. (Adapted from Fermentek)

3.2.4 *Metabolism of Aflatoxin*

Aflatoxin (Af) metabolism takes place in three steps:

1. Bioactivation
2. Conjugation
3. Deconjugation

1. **Bioactivation:** Toxic effects of aflatoxins are exerted in the first step. Aflatoxin B_1 is oxidized to different hydroxylated metabolites. The metabolic pathways comprise o-demethylation to Af P_1, reduction to aflatoxicol, and hydroxylation to Af B_1-8,9-epoxide (acutely toxic, mutagenic and carcinogenic), Af M_1 (acutely toxic), Af Q_1, or Af B_2 (both relatively toxic). Af B_1-8,9-epoxide is highly unstable; therefore, depending on the subsequent molecule, several reactions such as the formation of stable links between biological nucleophiles (RNA and DNA), point mutation, and DNA strand breaks take place. These reactions are associated with the carcinogenic effects of Af B_1 in humans and animals (Fig. 3.7).

   When water molecules are present, Af B_1-8,9-epoxide hydrolyses to Af B_1-8,9-dihydrodiol and becomes available to form bonds with serum proteins (lysine and albumin). This mechanism is responsible for the aflatoxins' toxicity (Biomin 2021b, http://www.mycotoxins.info/myco_info/science_moa.html).
2. **Conjugation:** With the help of the enzymes glutathione-*S*-transferase (GST), β-glucuronidase, and/or sulphate transferase, biotransformation of the compounds found in bioactivation step, Af P_1, aflatoxicol, Af B_1-8,9-epoxide, Af M_1, Af Q_1, or Af B_2 occurs in the second step yielding Af B_1-glutathione, Af B_1-glucuronide, and Af B_1-sulfate, respectively. Af B_1-glutathione is the major conjugate found. This conjugation is the principal detoxification pathway of activated Af B_1 in many mammals and is important in the prevention and reduction of Af B_1-induced carcinogenicity. The conjugates thus formed are easily excreted with the bile into the intestinal tract. The susceptibility of different animal species to Af B_1 carcinogenicity is inversely proportional to the cytosolic GST activity (Biomin 2021b, http://www.mycotoxins.info/myco_info/science_moa.html).
3. **Deconjugation:** The bacteria of the normal flora, *Bacteroides*, *Bifidobacterium*, *Escherichia coli*, etc., of the intestine are responsible for deconjugation. Deconjugation helps in the reabsorption of Af B_1 conjugates and their entry into the hepatic circulation, which is a part of the metabolic activity of the intestinal flora (Biomin 2021b, http://www.mycotoxins.info/myco_info/science_moa.html; Kenneth 2006).

3.2.5 Mechanism of Toxicological Damage

Aflatoxin binds to the guanine in deoxyribonucleic acid (DNA), inhibiting the signal for the formation of messenger ribonucleic acid (RNA). Interruption of protein synthesis leads to deficiencies of structural proteins, enzymes, and lipid acceptor protein. The long-term effects of impaired protein synthesis include hepatic steatosis and a variety of metabolic and functional imbalances (Zain 2011). The ill effects of aflatoxins on various systems are shown in Table 3.1 (Biomin 2021a, http://www.mycotoxins.info/en/mycotoxins/).

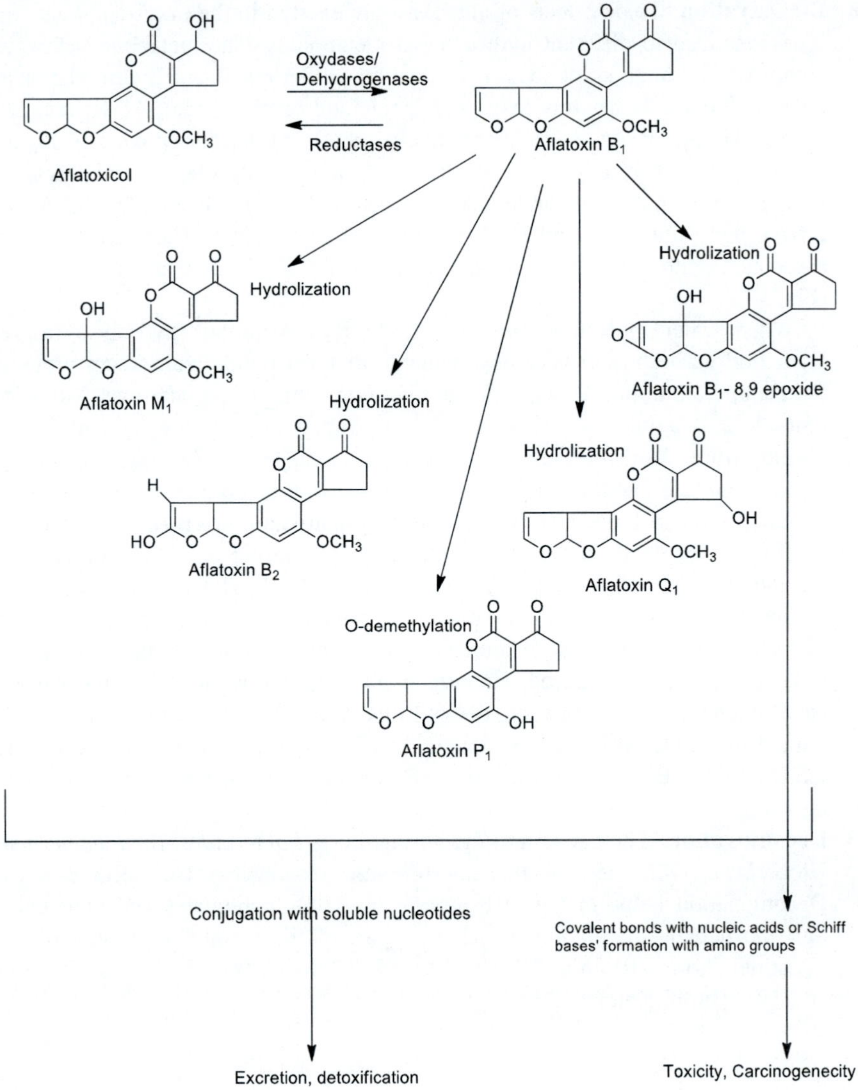

Fig. 3.7 Aflatoxin B_1 metabolism pathway

3.3 Fumonisins

Discovery of fumonisins occurred in 1988. The solubility of fumonisins in water accounts for their late discovery (Muro-Cach et al. 2004). Fumonisins are a collection of mycotoxins produced by *Fusarium*, Liseola section (Rheeder et al. 2002). Twelve analogues of fumonisin are known and the B (fumonisins B1, B2, and B3) series is the most significant one. Toxicologically, the most important *Fusarium*

Table 3.1 Detrimental effects of aflatoxins in humans and animals

Affected system	Effects/signs/symptoms
Genes/gene expression	Teratogenic effects: birth defects in the offspring
Genes/gene expression	Carcinogenic effects: higher incidence of cancer in exposed animals
Pathological changes	Weight variation of the internal organs (liver, spleen, kidneys enlargement, fatty liver syndrome), bursa of Fabricius and thymus reduction, change in the texture and colouration of the organs (liver, gizzard)
Circulatory system	Haematopoietic effects (haemorrhages, anaemia)
Immune system	Immunosuppression (decreased resistance to environmental and microbial stress; increased susceptibility to diseases)
Nervous system	Nervous syndrome (e.g. abnormal behaviour)
Skin	Dermatoxic effects (impaired feathering)
Urinary system	Kidney inflammation
Digestive system	Impaired rumen function, with decreased digestion of cellulose, decreased formation of volatile fatty acid, lowered proteolysis, decreased motility of rumen, diarrhoea
Reproductive system	Decreased breeding efficiency (birth of smaller and unhealthy offspring)

mycotoxins are deoxynivalenol (DON), trichothecenes, and T-2 toxin (T-2), zearalenone (ZEN), and fumonisin B_1 (FB_1) (Antonissen et al. 2014). Fumonisin B_1 is the most extensively studied member. They are not readily absorbed in the digestive tract and thus are distributed and removed from the body rapidly (Muro-Cach et al. 2004).

3.3.1 *Sources*

Fumonisins are mostly synthesized by *Fusarium moniliforme* and *F. proliferatem* (Carlson and Ensley 2003). Other species of Fusarium such as *F. napiforme*, *F. dlamini*, and *F. nygamai*, along with *Alternaria alternata* f. sp. lycopersici, also produce fumonisins (Bennett and Klich 2003; EFSA 2005). They are primarily found in maize but are also prevalent in wheat, sorghum, millet, and their agricultural products (Nair 1998).

3.3.2 *Physical and Chemical Properties*

Fumonisin B_1 is a white to off-white powder in appearance and is chemically known as 1,2,3-propanetricarboxylic acid, 1,1′-[1-(12-amino-4,9,11-trihydroxy-2-

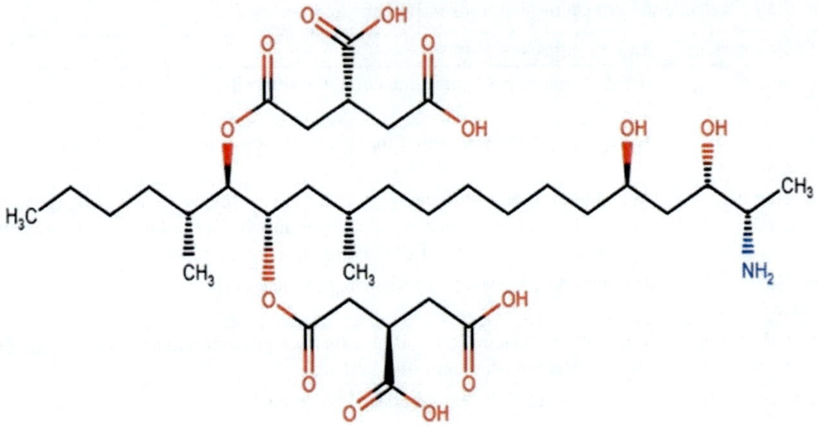

Fig. 3.8 Structure of fumonisin B1. (Adapted from Fermentek)

methyltridecyl)-2-(1-methylpentyl)-1,2-ethanediyl] ester (EFSA 2005; Marin et al. 2013) (Fig. 3.8).

3.3.3 Environmental Conditions Favouring the Production of Fumonisins

F. moniliforme and *F. proliferatem* are associated with temperate cereals since they require lower temperatures for growth and mycotoxin production as compared to *Aspergillus species* (Placinta et al. 1999). *F. verticillioides* and *F. proliferatum* are fumonisin-producing fungi that grow over a broad range of temperatures but only at comparatively high water activities ($a_w > 0.9$). The most important mycotoxins found in maize are fumonisins, particularly when grown in warmer regions. Fumonisin B_1, B_2, and B_3 (FBs) are formed in maize prior to harvest or during the early stages of storage. The concentration of FBs does not increase during storage except under extreme conditions. They are fairly heat-stable, and their toxin content is significantly reduced only during processes in which the temperature surpasses 150 °C. There is little degradation of FBs during fermentation (Marin et al. 2013).

3.3.4 Toxicity

Fumonisins (B_1 and B_2) are cancer-promoting metabolites that have a long-chain hydrocarbon unit (similar to that of sphingosine and sphinganine). Fumonisin contaminated feed causes a variety of species-specific pathologies like leukoencephalomalacia (hole in the head syndrome) in equine and rabbits and

Table 3.2 Detrimental effects of fumonisins on humans and animals

Affected system	Effects/signs/symptoms
Immune system	Immunosuppression (decreased resistance to environmental and microbial stresses; increased susceptibility to diseases)
Digestive system	Gastrointestinal effects (diarrhoea)
Circulatory system	Hematopoietic effects (haematological disorders; increased concentration of haemoglobin)
Nervous system	Neurotoxic effects
Digestive system	Hepatotoxic effects
Pathological changes	Kidneys and liver weight increase; liver necrosis; pancreatic necroses

pulmonary oedema and hydrothorax in swine (Muro-Cach et al. 2004). In the liver of rats, they turn out to be hepatotoxic, carcinogenic, and apoptotic (Hussein and Brasel 2001). Fumonisin B_1 (FB_1) is the most toxic amongst all other fumonisins. Toxic effects of fumonisins on humans and animals are given in Table 3.2 (Biomin 2021a, http://www.mycotoxins.info/en/mycotoxins/).

3.3.5 *Mechanism of Toxicological Damage*

The structural similarity of fumonisins with sphingoid bases, sphingosine, and sphingamine renders to its toxicity. Sphingosine is an important constituent of sphingolipids, which are vital for the membrane and lipoprotein structures as well as for cell regulations and communication (second messenger for growth and growth factors). Fumonisins disrupt the lipid metabolism pathway as they are inhibitors of sphinganine (sphingosine) *N*-acetyltransferase (ceramide synthase), which is an important enzyme for lipid metabolism (Biomin 2021a, http://www.mycotoxins.info/en/mycotoxins/).

The disruption of the pathway leads to the increase of many bioactive compounds and the decrease of others by various mechanisms such as the following:

1. Rapid increase in sphinganine (sometimes sphingosine)
2. Increase of sphingosine degradation products like sphinganine 1-phosphate
3. Decrease of complex sphingolipids

The defence mechanisms of the intestine such as cell proliferation, epithelial integrity, mucus layer, immunoglobulins (Ig), and cytokine production can be changed by various *Fusarium* mycotoxins.

The entry of foreign antigens including food proteins, xenobiotics (such as toxins and drugs), commensal microorganisms, and pathogens into the underlying tissues are prevented by the intestinal mucosa by acting as a barrier against them. *Fusarium* mycotoxins can affect the mucosal immunity, which comprises the innate and the adaptive immune system (Fig. 3.9). The tight junctions interconnect the intestinal

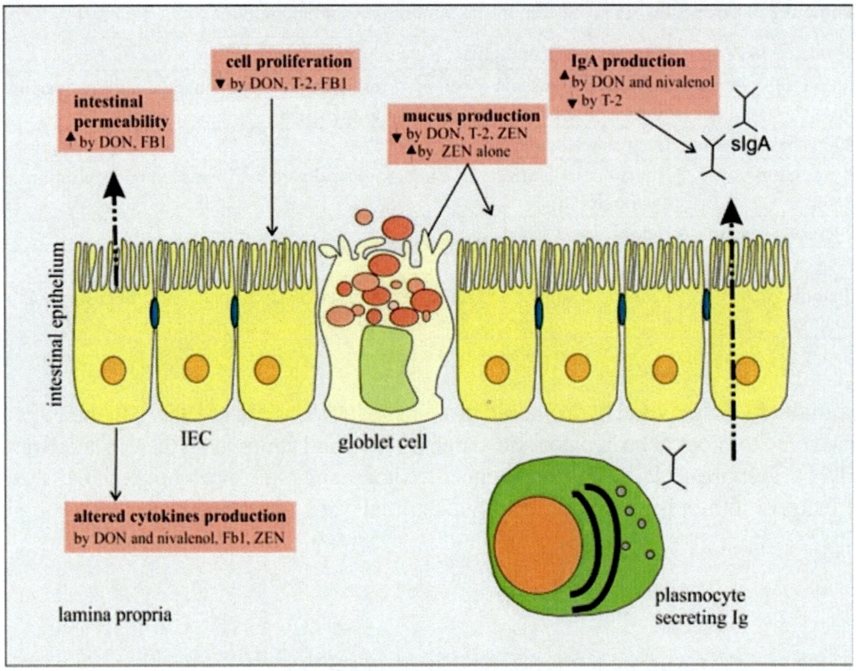

Fig. 3.9 The effect of *Fusarium* mycotoxins on the intestinal epithelium (IEC, intestinal epithelial cell). (This figure is reprinted in a modified form with permission from Antonissen et al. 2014)

epithelial cells, which are coated with mucus, secreted by goblet cells forming a major part of the innate immune system. By measuring the transepithelial electrical resistance (TEER), it has been shown that FB_1 and DON increase the permeability of the intestinal epithelial layer of human, porcine, and avian origin, as observed in several in vitro and ex vivo studies. *Fusarium* mycotoxins are able to negatively affect the proliferation and viability of human and animal intestinal epithelial cells (Antonissen et al. 2014).

3.4 Trichothecene

Trichothecenes are produced by various species such as the *Fusarium*, *Stachybotrys*, *Trichoderma*, *Cephalosporium*, *Trichothecium*, *Verticimonosporium*, and *Myrothecium*. They are a very large family of chemically related mycotoxins. On the basis of their producers and chemical characteristics, trichothecenes have been grouped into four classes:

1. T-2 toxin type A
2. Nivalenol type B

3. Crotocin type C
4. Macrocyclic type D

Type B trichothecenes have a carbonyl group at the eighth carbon position, while type A trichothecenes are not oxidized, hydroxylated, or esterified at this position (Döll and Dänicke 2011).

3.4.1 T-2 Toxin

3.4.1.1 Sources

T-2 toxins are predominantly found in grains, such as barley, maize, wheat, rice, soybeans, and particularly in oats and their products (Adhikari et al. 2017). *Fusarium moniliforme*, *F. avenaceum*, *F. culmorum*, *F. equiseti*, *F. graminearum*, *F. solani*, *F. nivale*, *F. poae*, *F. roseum*, *F. tricinctum*, *F. sporotrichiella*, and other fungal species are associated with the production of T-2 toxin (Adhikari et al. 2017).

3.4.1.2 Environmental Conditions Favouring the Production of T-2 Toxin

Tropical conditions like high temperatures, moisture levels, monsoons, unseasonal rains during harvests, and flash floods enhance the growth of mycotoxigenic fungi. Growth of *Fusarium* sp. has been reported in cereals that are cultivated in the regions of humid subtropical climates.

3.4.1.3 Chemical Properties

Trichothecenes (T-2) belong to a group of closely related compounds called sesquiterpenoids (Fig. 3.10). There are around 150 derivatives of trichothecenes studied so far. T-2 toxin is the most extensively studied mycotoxin because of its high toxicity and abundance. In diverse environments, they resist degradation in conditions of different temperatures and light levels, but they can be easily deactivated by strongly alkaline or acidic conditions. Trichothecenes are non-volatile and low-molecular-weight (MW 250–550) compounds. They are relatively soluble in water but highly soluble in acetone, dimethyl sulfoxide (DMSO), chloroform, ethyl acetate, ethanol, methanol, and propylene glycol. Usually, purified trichothecenes have a low vapour pressure but vaporize when heated in organic solvents (Wilson et al. 2003).

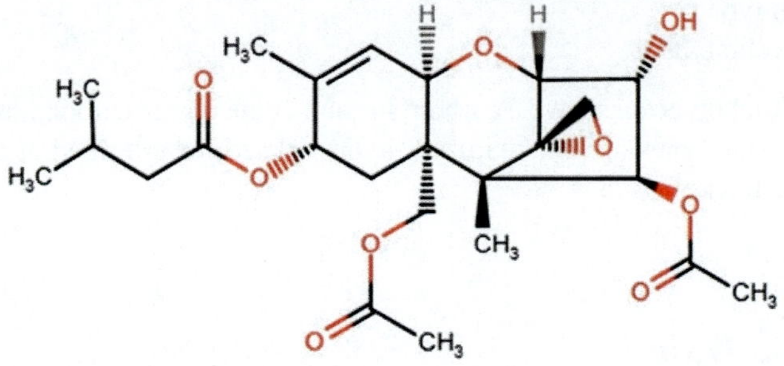

Fig. 3.10 Structure of T2 toxin. (Adapted from Fermentek)

3.4.1.4 Toxicity

Eukaryotic microorganisms, plants, many invertebrates, fishes, birds, humans, and other mammals are affected by trichothecenes. Their toxicity differs depending on animal species. Trichothecenes affect the rapidly proliferating cells once they enter the systemic circulation.

3.4.1.5 Mechanism of Action

Trichothecene mycotoxins are cytotoxic to most of the eukaryotic cells (Wannemacher et al. 1997). They are powerful inhibitors of protein, RNA, and DNA syntheses and can interact with the cell membrane (Biomin 2021a, http://www.mycotoxins.info/en/mycotoxins/). Trichothecenes bind to active polysomes and ribosomes, interrupt peptide linkages, reduce initiation and termination sequences, and disrupt the ribosomal cycle. Their toxicity is primarily based on the 12,13-epoxytricothecene ring. Trichothecenes may act

1. Either by inhibition of the initial step of protein synthesis (e.g. T2 toxin, HT2 toxin, diacetoxyscirpenol)
2. Or by inhibition of the elongation termination step (e.g. deoxynivalenol)

3.4.2 *Deoxynivalenol*

Deoxynivalenol is the most commonly detected *Fusarium* mycotoxin (Akande et al. 2006).

3.4.2.1 Sources

Deoxynivalenol (DON) is one of the most abundant contaminants in maize and wheat worldwide and is chiefly produced by *F. culmorum* and *F. graminearum.*

These species also contaminate barley, oats, sorghum, rye, safflower seeds, and mixed feeds (site.iugaza.edu.ps/akichaoi/files/2010/02/Mycotoxin.ppt.ppt).

3.4.2.2 Chemical Properties

Deoxynivalenol is a major type B trichothecene mycotoxin. Chemically, DON is described as 12, 13-epoxy-3,7,15-trihydroxytrichotec-9-en-8-one (Fig. 3.11). It crystallizes as colourless needles. It is stable at high temperatures, i.e. at 120 °C, and moderately stable at 180 °C. It is soluble in water and other polar solvents like acetonitrile, aqueous methanol, and ethyl acetate (Döll and Dänicke 2011).

3.4.2.3 Environmental Conditions Favouring the Production of DON

Wet and cold weather during the maturation of grains make them especially susceptible to *F. graminearum* infection, which causes the so-called "scabby wheat" and simultaneously produces the toxin. The optimal temperature for DON production is about 24 °C. Inadequate storage of grains may lead to the production DON (Wu et al. 2017).

3.4.2.4 Toxicity

A number of toxic effects like feed refusal, weight loss, and vomiting are observed in acute and chronic ingestion of DON by humans and animals (site.iugaza.edu.ps/akichaoi/files/2010/02/Mycotoxin.ppt.ppt). Feed refusal in animals caused by DON is a learned response known as taste aversion (http://www.msdvetmanual.com/toxicology/mycotoxicoses/trichothecene-toxicosis). Other signs of the disease are restricted since denial to eat contaminated feedstuffs is the typical sign. Due to the non-availability of other food, animals may eat the contaminated feed reluctantly. In some cases, excessive salivation and vomiting might also take place. In the past, the capacity to cause vomiting had been attributed to DON only, and therefore, its common name is vomitoxin. Conversely, other members of the family trichothecene can also induce vomiting. Moreover, DON toxicity also results in reduced milk production in dairy cattle and has been reported to inhibit reproductive performance

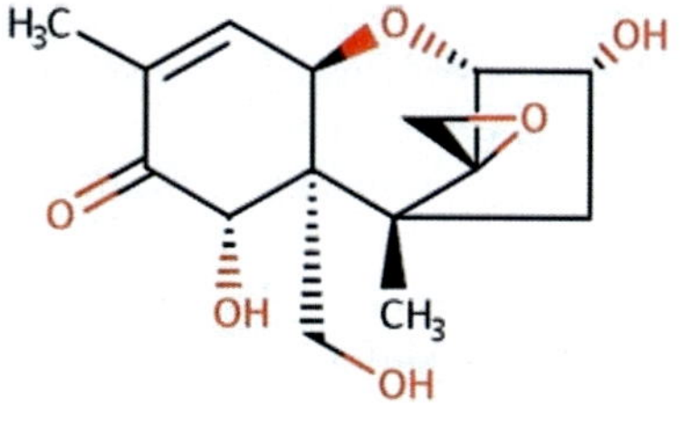

Fig. 3.11 Structure of DON. (Adapted from Fermentek)

Table 3.3 Detrimental effects of trichothecenes on humans and animals

Affected system	Effects/signs/symptoms
Circulatory system	Haematopoietic effects (haemorrhages, blood pattern disorders)
Immune system	Immunosuppression (decreased resistance to environmental and microbial stress, increased susceptibility to diseases)
Digestive system	Gastrointestinal effects (gastroenteritis/inflammation of the rumen, vomiting, feed refusal)
Reproductive system	Decreased breeding efficiency (birth of smaller and unhealthy offspring)
Nervous system	Neurotoxic effects (restlessness, lack of reflexes, abnormal wing positioning, nervous syndrome)
Skin	Dermotoxicity (oral and dermal lesions, necrosis)
Pathological changes	Necrosis of the lymphoid and haematopoietic tissues, gizzard lesions

in several animal species. Immunosuppression and immunostimulation are also variable effects of DON, as suggested by experimental studies (Akande et al. 2006).

Most trichothecenes are immunosuppressors. Toxicologically, DON induces anorexia and emesis both in humans and animals. DON is also a hyper inducer of cytokines. Swine are most susceptible to DON-contaminated feed (site.iugaza.edu.ps/akichaoi/files/2010/02/Mycotoxin.ppt.ppt). The toxic effects of trichothecenes on humans and animals are listed in Table 3.3 (Biomin 2021a, http://www.mycotoxins.info/en/mycotoxins/).

3.5 Ergot Alkaloids

A fascinating group of fungal metabolites is ergot alkaloids (EA) (Bennett and Klich 2003). *Claviceps purpurea* and some *Acremonium* species produce ergot alkaloids like ergotamine, ergocristine, and ergostine that cause sleepy grass toxicosis in farm animals (Hussein and Brasel 2001).

3.5.1 Sources

All species of the genus *Claviceps* (class: *Ascomycetes*, family: *Clavicipitaceae*) are rich sources of ergot alkaloids (EA). Endophytic clavicipitaeous fungi of the genera *Epichloe* (*Acremonium*) and *Balansia*, as well as epiphytic *Atkinsonella*, produce EA. Like the parasitic *Claviceps*, these fungi are found exclusively on grasses but have various host relationships ranging from mutualism to parasitism. Other distantly related fungi such as *Penicillium* and *Aspergillus* are also reported to produce EA.

3.5.2 Chemical Properties

They are derived from a tetracyclic ergoline ring system and are classified as indole alkaloids (Akande et al. 2006). Ergot alkaloids are structurally related to the hallucinogenic drug known as lysergic acid diethylamide (Bennett and Klich 2003) (Fig. 3.12).

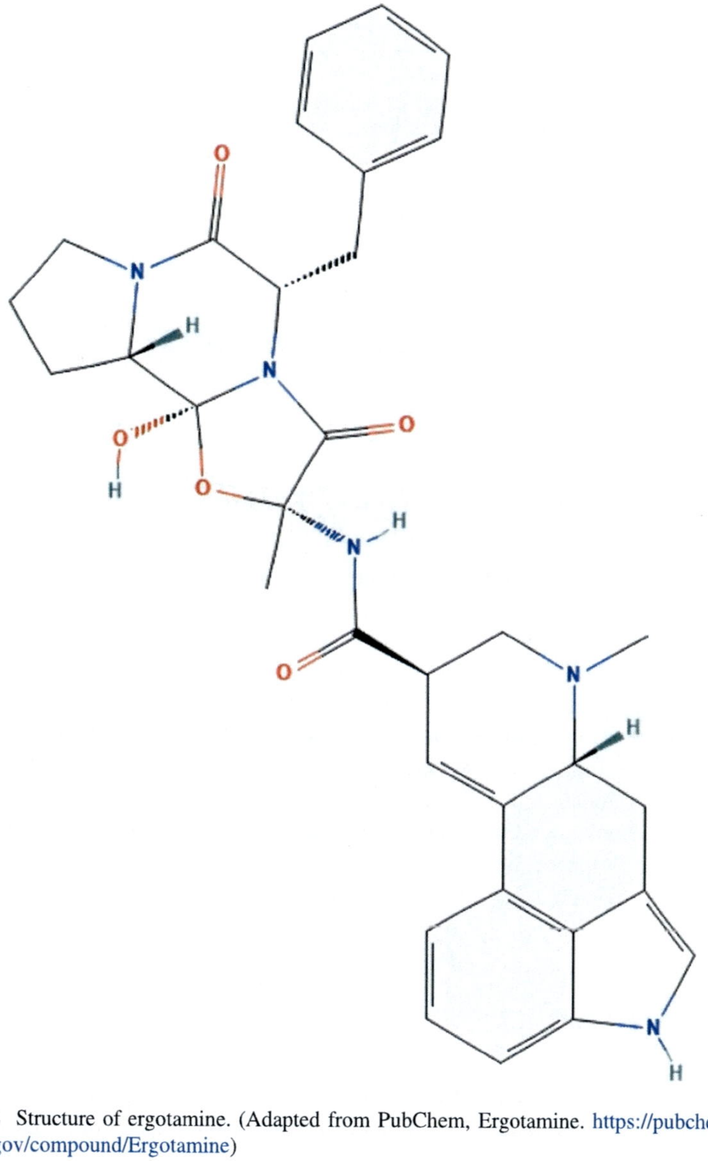

Fig. 3.12 Structure of ergotamine. (Adapted from PubChem, Ergotamine. https://pubchem.ncbi.nlm.nih.gov/compound/Ergotamine)

Table 3.4 Detrimental effects of ergot alkaloids on humans and animals

Affected system	Effects/signs/symptoms
Nervous system	Depression of the CNS, convulsions, ataxia
Endocrine system	Inhibit the pituitary release of prolactin and stoppage of mammary development in late gestation as well as delayed initiation of milk secretion, leading to agalactia at parturition
Vascular system	Vasoconstriction, vascular endothelium injury
Immune system	Hypersensitivity
Reproduction system	Abortion
Genes/gene expression	Teratogenic effects
Pathological changes	Gangrene and loss of limbs

3.5.3 *Toxicity*

There are two kinds of ergotism known in humans and animals: convulsive and gangrenous (Hussein and Brasel 2001). Table 3.4 states the toxic effects of ergotamine in different systems of humans and animals.

3.5.4 *Mechanism of Toxicological Action*

By direct action on the muscles of the arterioles, ergot causes vasoconstriction while repeated doses result in vascular endothelium injury. Initially, these actions reduce blood flow and ultimately lead to complete stasis with terminal necrosis of the extremities due to thrombosis. The extremities are predisposed to gangrene in a cold environment. Stimulation and depression of the central nervous system (CNS) are also an effect of ergot. In many mammalian species, ergot alkaloids inhibit the pituitary release of prolactin and stoppage of mammary development in late gestation. It also causes delayed initiation of milk secretion, leading to agalactia at parturition. Analogous to the "summer syndrome" caused by fescue toxicosis, ergot alkaloids have also been related to dyspnoea, heat intolerance, and reduced milk production in dairy cattle (http://www.merckvetmanual.com/mvm/zk/toxicology/mycotoxicoses/ergotism.html).

3.6 Ochratoxin

In 1965, an *Aspergillus ochraceus* metabolite known as Ochratoxin A was discovered during the screening of fungal metabolites for the presence of mycotoxins. Soon after that, it was isolated in the United States from a commercial corn sample and classified as a powerful nephrotoxic compound (Bennett and Klich 2003).

3.6.1 Sources

Aspergillus alliaceus, *A. niger*, *A. glaucus*, *A. melleus*, *A. carbonarius*, and *A. auricomus* are known to produce mycotoxins belonging to the family ochratoxin. It is also produced by a common contaminant of barley, *Penicillium verrucosum* (Chu and Wilson 1973; Ciegler et al. 1973; Pitt 1987; Abarca et al. 1994; Bayman et al. 2002). Similarly, *Fusarium* species also produce ochratoxin (Bennett and Klich 2003; Richard 2007). Ochratoxin A is present in barley, oats, rye, wheat, coffee beans, and other plant products, barley having particularly high likelihood of contamination. Ochratoxin is present in some wines, especially those from grapes contaminated with *A. carbonarius* (Haackgeb 1978; Marquardt and Frohlich 1992; Van Egmond and Speijers 1994; Pitt 2000).

3.6.2 Chemical Structure

Ochratoxin is a derivative of dihydromethyl isocoumarin and is linked with an amide bond to the amino group of L-β-phenylalanine (Pubchem, Ochratoxin A Compound) (Fig. 3.13).

3.6.3 Toxicity

Apart from being a nephrotoxin, the studies conducted on animals indicate that ochratoxin A is hepatotoxic, immune-suppressant, a powerful teratogen, and also a carcinogen (Kuiper-Goodman and Scott 1989; Beardall 1994). Various effects of ochratoxin A are listed in Table 3.5 (Biomin 2021a, http://www.mycotoxins.info/en/mycotoxins/).

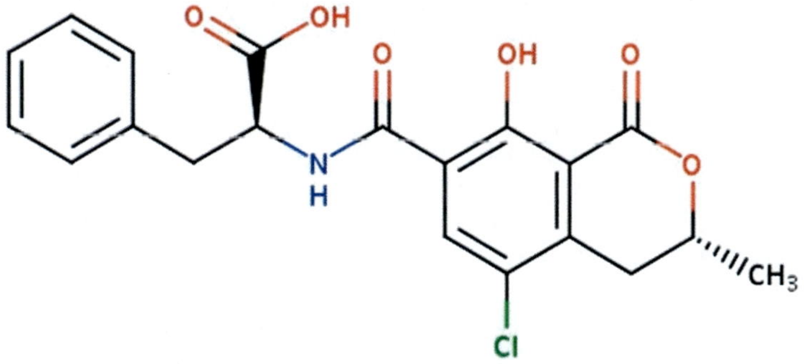

Fig. 3.13 Structure of ochratoxin A. (Adapted from Fermentek)

Table 3.5 Detrimental effects of ochratoxins on humans and animals

Affected system	Effects/signs/symptoms
Circulatory system	Haematopoietic effects (haematological disorders, haematuria, and blood in faeces)
Nephrotoxic effects	Increased water consumption, kidney and bladder dysfunction
Immune system	Immunosuppression (decreased resistance to environmental and microbial stress, increased susceptibility to diseases)
Digestive system	Gastrointestinal effects (diarrhoea), hepatotoxic effects (liver damage)

3.6.4 *Mechanism of Toxicological Action*

The physiological functions of a cell are disrupted by ochratoxin A in a number of ways, but it plays a key role in disrupting the phenylalanine metabolism, usually by obstructing the enzyme engaged in the synthesis of the phenylalanine tRNA complex (Bunge et al. 1979; Marquardt and Frohlich 1992). Ochratoxins might also interact with other enzymes that use phenylalanine as a substrate, such as phenylalanine hydroxylase, which catalyses the irreversible hydroxylation of phenylalanine to tyrosine. Moreover, it alters the mitochondrial membrane transportation system, inhibits ATP production, and enhances membrane lipid peroxidation and superoxide and hydrogen peroxide radical formation (Biomin 2021a, http://www.mycotoxins.info/en/mycotoxins/). Blood, tissues, and milk (including human milk) show the presence of ochratoxin (Marquardt and Frohlich 1992).

3.6.5 *Metabolism of OTA*

The major metabolite of ochratoxin is ochratoxin A, a hydrolysis product by the gut microflora without the phenylalanine moiety. The other hydroxylated derivatives of ochratoxin are 4(R), 4(S), and 10 OH ochratoxin (Heidler and Schatzmayr 2003) (Fig. 3.14).

3.6.5.1 Absorption/Excretion/Residues

Around 40 and 66% of ochratoxin is absorbed from the gastrointestinal tract depending on the animal species. The small intestine has been shown to be the major site of absorption. The maximum absorption of OTA occurs in the jejunum. Ochratoxins are distributed in the blood by rapidly binding to the serum albumin. Since the toxin has a high affinity for serum protein, it has a long biological half-life. Therefore, the ochratoxin protein adduct in serum can be used as a biomarker for ochratoxins exposure. It primarily accumulates in the kidneys, followed by the liver,

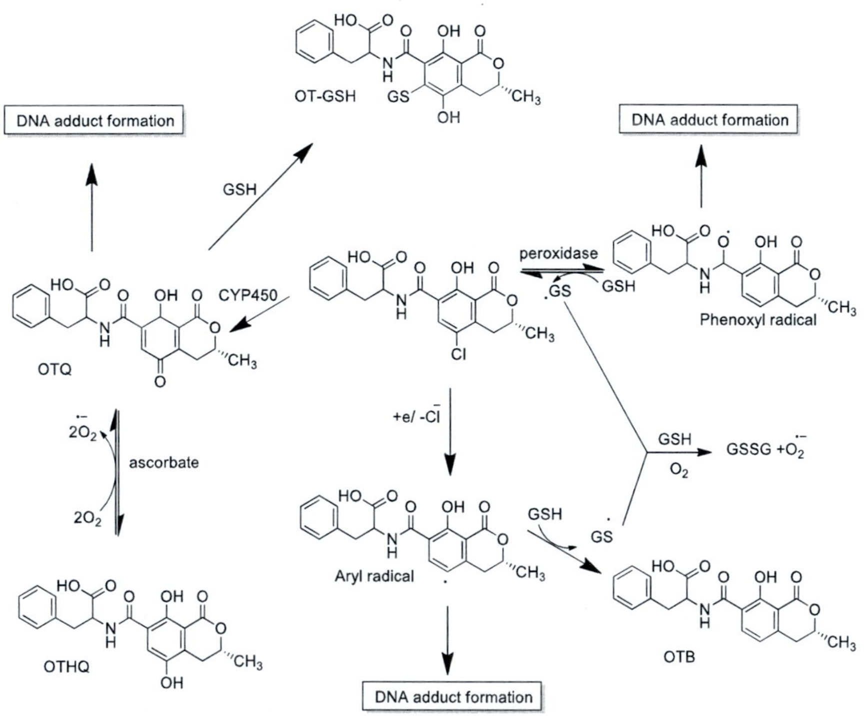

Fig. 3.14 Bioactivation of ochratoxin A producing genotoxic metabolites (*OTA* ochratoxin A, *OT-GSH* OTA-glutathione conjugate, *OTQ* OTA-quinone, *OTHQ* OTA-hydroquinone, *OTB* ochratoxin B)

muscle, and fat. Due to this accumulation, the toxin may potentially be carried over into meat (Binder 2007; Kőszegi and Poór 2016).

3.7 Zearalenone

Zearalenone is a mycotoxin having toxicity much lower than the other well-known mycotoxins. A more appropriate categorization for this compound would be as a non-steroidal oestrogen or mycoestrogen because of its oestrogen-like properties, which have been reported in swine (Kőszegi and Poór 2016).

3.7.1 Sources

Zearalenone (ZEN) is a mycotoxin produced by several species of *Fusarium* such as *F. graminearum*, *F. culmorum*, *F. cerealis*, *F. equiseti*, *F. verticillioides*,

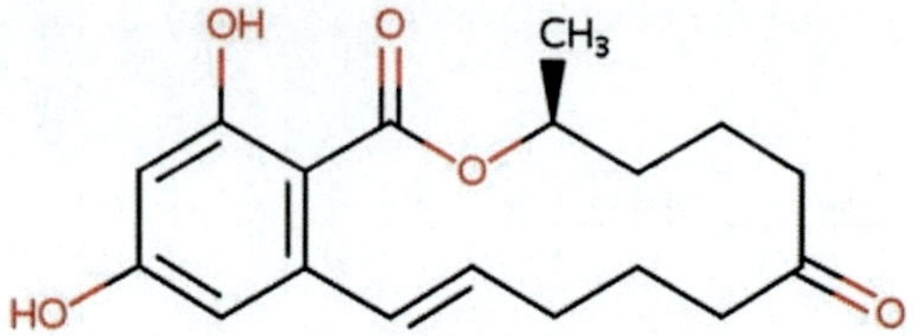

Fig. 3.15 Structure of zearalenone. (Adapted from Fermentek)

F. crookwellense, *F. semitectum*, and *F. incarnatum*, *F. graminearum* being its major producer (Marin et al. 2013). Fungi-producing ZEN contaminates corn and also colonizes barley, millet, oats, rice, sorghum, and wheat. In addition, the toxin has been detected in cereal products like flour, soybeans, malt, and beer.

3.7.2 Chemical Properties

Zearalenone (6-[10-hydroxy-6-oxo-*trans*-1-undecenyl]-B-resorcyclic acid lactone), a secondary metabolite from *Fusarium graminearum* (teleomorph *Gibberella zeae*), was given the trivial name zearalenone because it was produced by *G. zeae*, and it was a mixture of resorcylic acid lactone, -ene (due to the presence of the C-1 to C-2 double bond), and -one, due to the presence of C-6 ketone (Urry et al. 1966) (Fig. 3.15). Only high temperatures and alkaline conditions can cause its degradation. However, ZEN is heat-stable up to 150 °C (Ryu et al. 1999).

3.7.3 Environmental Conditions Favouring the Production of Zearalenone

ZEN is produced by different species of *Fusarium* fungi, common soil fungi, in temperate and warm countries. They regularly contaminate cereal crops worldwide (Ismaiel and Papenbrock 2015). Usually, the *Fusarium* species colonize crops in moist and cold field conditions before harvesting, but growth and toxin production may also occur post-harvest if the crop is not handled and dried properly and under poor storage conditions (EFSA 2011).

3.7.4 Toxicity

Zearalenone at a concentration of as low as 1.0 ppm may result in hyperoestrogenic syndromes in pigs, while at higher levels, it can cause disrupted conception, abortion, and other issues. Reproductive problems caused by ZEN have also been observed in cattle and sheep and rats (El-Nezami et al. 2002). In rats, they are

Table 3.6 Detrimental effects of zearalenone on humans and animals

Affected system	Effects/signs/symptoms
Digestive system	Gastrointestinal effects (diarrhoea)
Reproduction system	Reproductive effects (feminization, enlargement of mammary glands, impaired semen quality, testicular atrophy, swollen prepuce)
Genes/gene expression	Teratogenic effects (splay legs: impaired functionality of the hind leg muscles in piglets soon after their birth, resulting in an inability to stand or walk) (Papatsiros 2012)
Pathological changes	Atrophy of ovaries, uterus hypertrophy

known to cause sterility syndrome in the offspring as manifested by the administration of androgen or oestrogen (Kuiper-Goodman and Scott 1989). Table 3.6 lists the systems in humans and animals affected by zearalenone.

3.8 Patulin

Patulin (PAT) is predominantly produced by *Penicillium patulum* and *Penicillium expansum* and several species of *Penicillium* as well as other moulds like *Aspergillus* spp. Patulin can also be produced in apples, grapes, pears, peaches, and apricots infected by *Paecilomyces variotii*, *Byssochlamys nivea*, and *Byssochlamys fulva* (Bokhari et al. 2009; Pal et al. 2017). PAT was discovered in the mid-1900s as an antibiotic and anticancer drug and was used as a treatment for the common cold until 1950. The clinical trials of patulin (under the brand name tericinin) were performed by a British company (Chalmers and Clarke 2004; Puel et al. 2010). However, in the 1960s, due to its toxicological aspect, PAT was designated as a mycotoxin. PAT-producing fungi are generally post-harvest contaminants of vegetables, fruits, animal feeds, and residues of wheat straw. Patulin is a heat-stable compound and can withstand various food-processing methods and can remain in jam, jelly, and other fruit-derived products (Saleh and Goktepe 2019; Yuan et al. 2010). The contamination of PAT, particularly in apples and apple-based products, has been reported worldwide (Cheraghali et al. 2005; Murillo-Arbizu et al. 2009). Exposure of patulin leads to immunological, neurological, and gastrointestinal effects. Evaluation of the health hazards due to patulin consumption by humans has led many countries to legalize their quantity in food (Puel et al. 2010).

3.8.1 Sources

The presence of patulin has been reported in apples and apple products such as juices, jams, and ciders. It has also been detected in other fruits like apricots, bananas, blueberries, cherries, grapes, peaches, pears, plums, and strawberries. High levels of patulin have also been found in different grains (barley, corn, and wheat) and their processed products, shellfish, and vegetables (Perishables Handling Quarterly 1997; Llewellyn et al. 1998; Fernández-Cruz et al. 2010). It is produced by a large number of fungi of the genera *Byssochlamys*, *Eupenicillium*, *Penicillium*, *Aspergillus*, and *Paecilomyces* (Ismaiel and Papenbrock 2015).

The patulin producing fungal species are *P. expansum*, *P. leucopus*, *P. vulpinum*, *P. claviforme*, *P. griseofulvum*, *P. patulum*, *P. carneum*, *P. clavigerum*, *P. concentricum*, *P. coprobium*, *P. dipodomyicola*, *P. glandicola*, *P. gladioli*, *P. marinum*, *P. paneum*, *P. sclerotigenum*, *P. cyclopium*, *P. granulatum*, *P. divergens*, *P. melinii*, *P. lapidosum*, *P. aurantiogriseum*, *P. canescens*, *P. chrysogenum*, *P. roqueforti*, *P. variabile*, *P. purpurogenum*, *Aspergillus clavatus*, *A. giganteus*, *A. longivesica*, *A. terreus*, *Byssochlamys fulva*, *B. nivea*, and *Paecilomyces saturates* (Fernández-Cruz et al. 2010; Ismaiel and Papenbrock 2015).

3.8.2 Environmental Conditions Favouring the Production of Patulin

P. expansum produces patulin within a temperature range of 0–24 °C. The minimum water activity (a_w) for patulin production is 0.99 (Fernández-Cruz et al. 2010). It is mainly produced in rotten parts of the fruits after infestation with PAT-producing fungus that can occur at 0 °C during storage or develops rapidly when fruits are returned to room temperature (Sanderson and Spotts 1995). The level of patulin also tends to increase when kept at room temperature (20 °C) for short periods of time during the processing of fruits (Ismaiel and Papenbrock 2015).

3.8.3 Physical and Chemical Properties

Patulin is crystalline and colourless, has a low-pH, and is soluble in organic solvents and water as well. The IUPAC name of PAT is 4-hydroxy-4H-furo[3,2-c]pyran-2(6H)-one). The molecular formula of PAT is $C_7H_6O_4$ (Fig. 3.16) while the molar mass is 154.12. Chemically, it is a polyketide lactone, is not destroyed by pasteurization or thermal denaturation, and thus is heat-stable, but its stability is reduced

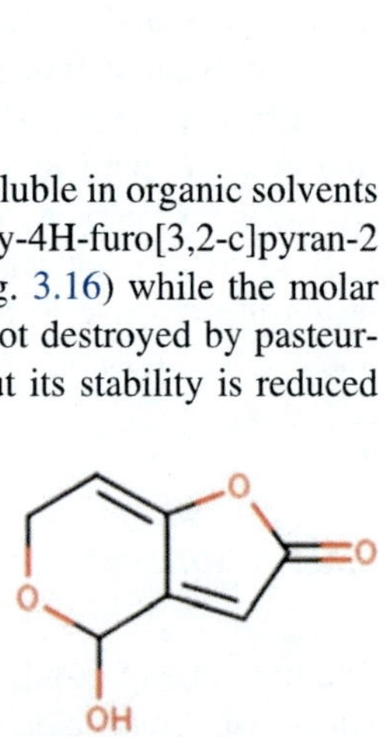

Fig. 3.16 Structure of patulin. (Adapted from ChemSpider)

after fermentation (http://www.sigmaaldrich.com/catalog/product/sigma/p1639?lang=en&region=US). The biosynthesis of patulin takes place in the presence of isoepoxydon dehydrogenase (IDH), an important enzyme present in the fungi that produce the toxin. Its biosynthesis is also controlled by nitrogen, manganese, pH, and the abundancy of necessary enzymes (Puel et al. 2010). As it reacts with sulphur dioxide, antioxidant and antimicrobial agents can be used for its destruction. PAT is considered the best-characterized mycotoxin until now.

3.8.4 Toxicity

The affinity of patulin towards sulfhydryl groups (SH) results in the inhibition of enzymes, leading to its toxic nature. Its oral LD_{50} in rodent models ranges between 20 and 100 mg/kg, while in poultry, it ranges between 50 and 170 mg/kg (Puel et al. 2010). The oral route of exposure is less toxic than other routes. Neurotoxicity (i.e. convulsions), pulmonary congestion, gastrointestinal problems, and oedema are the major acute toxicity manifestations of patulin (McKinley et al. 1982). The effects of patulin on humans and animals are shown in Table 3.7.

The consumption of fruits and their products contaminated with PAT leads to acute and sub-acute toxicities in humans and animals. Patulin interferes in DNA synthesis, thus causing genotoxicity in animals. It was listed as a genotoxic compound by WHO (1998). The oral administration of PAT induces less toxicity than intra-peritoneal and intravenous administrations (WHO 1998). As patulin is carcinogenic to animals only, it is classified as category 3 toxin by IARC (1993). However, its immunotoxicity is far more detrimental (Vidal et al. 2019). In humans, the sub-acute and chronic toxicity of PAT causes immunosuppressive and teratogenic effects (Mandappa et al. 2018). In animals, it has been found to cause haemorrhage and inflammation in the lungs and brain, convulsions, pulmonary congestion, and capillary damage in the spleen, liver, and kidney (Groopman and Kensler 2005). Patulin is immunosuppressive and can also induce GI tract malfunction and nephropathy. Vomiting, ulceration, and agitation are also manifested in PAT toxicity (Altomare et al. 2021). The LD_{50} values of PAT in rodents and in

Table 3.7 Detrimental effects of patulin on humans and animals

Affected system	Effects/signs/symptoms
Nervous system	Neurotoxicity (i.e. convulsions)
Respiratory system	Pulmonary congestion
Digestive system	Gastrointestinal problems (inflammation and ulceration of the stomach and intestine, epithelial cell degeneration, and haemorrhages)
Genes/gene expression	Teratogenic, carcinogenic, and mutagenic effects
Pathological changes	Oedema

Table 3.8 Toxicological profiles of PAT in humans, animals, and cell lines

Animal/cell line	Toxic effects
Zebra fish (*Danio rerio*)	Renal cell disarrangement (Wu et al. 2012)
Chicken	Renal malfunction; inhibits renal ATPases (Devaraj et al. 1986)
Chinese hamster	Teratogenic; formation of defective embryos with neural anomalies (Smith et al. 1993)
– Lung fibroblast V79 cells	Chromosomal aberrations (structural) (Alves et al. 2000)
– Ovary cells	Chromosomal breakage (Liu et al. 2003)
Mouse	Cutaneous haemorrhage, kidney malfunction, haemorrhage in the brain and lung, elevation of p53, cytochrome C expression in renal cells (Becci et al. 1981; Roll et al. 1990; Boussabbeh et al. 2016)
Macrophages	Decreases ROS generation, increased in splenic T lymphocytes, suppression in serum Ig concentration and neutrophilia (Bourdiol et al. 1990; Puel et al. 2010)
– Carcinoma cell line (FM3A)	Gene mutations, chromosomal aberrations (Umeda et al. 1977)
– Lymphoma cells (L5178Y)	Gene mutations, chromosomal aberrations (Schumacher et al. 2005)
Swiss ICR mice	Convulsions, GI tract dilation, ulceratin, enterotoxaemia, dyspnoea (McKinley and Carlton 1980a, b)
Rat	Abortion
– Alveolar macrophages	Membrane function alteration, protein synthesis inhibition (Reddy et al. 1978)
Human	Reduction in cytokine secretion
– Promyelocytic leukaemia cells (HL-60)	Reduction in ROS (Liu et al. 2007)
– Circulating leucocytes	Genotoxic activity (Liu et al. 2003)
– Intestinal epithelial Caco-2 cells and hepatoma HepG2 cells	Cell toxicity (Saleh and Goktepe 2019; Ramalingam et al. 2019)

poultry birds were 29 and 70 mg/kg, respectively. The sub-acute toxicity of patulin has been found to induce neurotoxicity, renal malfunction, weight loss, and changes in the stomach and intestine of rats. PAT can inhibit the activity of various intestinal enzymes involved in lipid metabolism (Kharayat and Singh 2018). The 0.1 mg/kg b. w./day of patulin has been shown to cause interstitial tissue enlargement, lymphoid cell infiltration in thyroid, Leydig cell hyperplasia, and low sperm count in rats when administered orally for 90 days. The toxicological profiles of PAT in humans and animals are listed in Table 3.8.

3.8.5 Mechanism of Toxicological Action

Patulin disrupts the function of the plasma membrane in cultured LLC-PK1 renal cells by the inhibition of Na-K ATPase (Llewellyn et al. 1998). The affinity of patulin towards sulfhydryl groups such as cysteine or glutathione (SH) results in the inhibition of enzymes, leading to its toxic nature. It is a highly potent inhibitor of RNA polymerase (Riley and Showker 1991). Another important harmful effect of patulin is that it causes injury to the gastrointestinal tract when tested in vivo. It is also responsible for the inflammation and ulceration of the stomach and intestine (Riley and Showker 1991). The inflammatory response of the gastrointestinal tract might be due to the irritant nature of patulin. Studies show that when administered orally to rats, 35% of the patulin was recovered in faeces and 35% in urine within 24 h of administration, indicating absorption of mycotoxin in the intestine (Dailey et al. 1977).

3.9 Alternariol (AOH) and Alternariol Monomethyl Ether (AME)

Legislation in various countries is available for well-established mycotoxins such as aflatoxins (AF), ochratoxin A (OTA), and trichothecenes in a variety of fruits and cereals. However, information regarding *Alternaria* mycotoxins in fresh fruit, vegetables, and cereals and by-products is limited. As yet, there are no specific international regulations for *Alternaria* mycotoxins in food and feed (Alexander et al. 2011). Report of European Food Safety Authority (EFSA) on the risks of *Alternaria* toxins for animal and public health (EFSA 2011) summarized data on *Alternaria* toxins analysed in feed and agricultural commodities, based on the available data from the literature. According to EFSA's report, the contamination ratio of *Alternaria* toxins in food is alternariol (AOH) > alternariol monomethyl ether (AME) > tentoxin (TEN) > tenuazonic acid (TeA) > altertoxins (ALTs) > stemphyltoxin III >*Alternaria alternata* toxins (AAL toxins) including AAL-TA and AAL-TB. Therefore, the most studied *Alternaria* mycotoxins are AOH and its monomethyl ether derivative (AME) as well as TEN. The toxicological database on *Alternaria* toxins is still limited, and their risk assessment reports remain inconclusive (Pose et al. 2010).

3.9.1 Sources

Alternaria mycotoxins are metabolites produced by several species of *Alternaria* such as *A. arborescens* or *A. alternata*, which infest vegetable plants, fruits, oil crops, and cereals. Other species of *Alternaria* and *Phomopsis* and various strains of

Stagonospora nodorum have also been reported to produce AOH and AME (Scott and Stoltz 1980).

Occurrences of AOH, AME, and other *Alternaria* toxins naturally have been known in various fruits, processed fruit products such as apple juice, tomato products, wheat, sunflower seeds, oilseed rape meal, flax seed, linseed, and pecans. They have also been reported in Canadian lentils. Fruits such as strawberries might also have the presence of these mycotoxins. AOH and AME are also found in grains or grain-based foods (Juan et al. 2016).

3.9.2 Conditions Favourable for the Production of Alternaria Toxin

Mycotoxin production is remarkably affected by different environmental factors, mostly temperature and water activity (a_w). *A. arborescens* can grow in fruits during the storage period, which may extend for several weeks, being able to produce AOH, AME, and TeA, and frequently at high levels (Alexander et al. 2011). AOH and AME remain unaltered under moist baking conditions like in the baking of bread (therefore, they might be associated with grain-based foods of Canadian origin). Strawberries are usually stored in refrigerated chambers and eventually at room temperature (RT). At the range of 20–25 °C and high water activity of fruits ($a_w = 0.995$), *A. arborescens* can germinate in less than 1 day. Consequently, at warm storage temperatures, the risk of fungal contamination increases. Although the optimum growth temperatures reported for *Alternaria* ranged between 22 and 30 °C, minimum growth temperatures ranging from 2.5 to 6.5 °C, and even lower, from 0 to 5 °C, in cooler regions have also been reported (Ostry 2008). Thus, this microorganism is capable of growing at low temperatures, currently used during transport and harvest, and possibly that is the reason why a high incidence of *Alternaria* in fresh vegetables has been detected (Tournas and Katsoudas 2005). Even though refrigeration temperatures represent the best strategy to control the pathogen, extremely low temperatures are not recommended for storing fresh fruits as it may lead to a significant reduction in sensorial quality, flavour, and colour development.

3.9.3 Chemical Structure

Alternariol (AOH) and alternariol monomethyl ether (AME) are benzopyrone mycotoxins produced by *Alternaria alternata* (Figs. 3.17 and 3.18). They are structurally characterized as 3,7,9-trihydroxy-1-methyl-6H-dibenzo [b,d] pyran-6-one and 3,7-dihydroxy-9-methoxy-1-methyl-6H-dibenzo pyran-6-one, respectively.

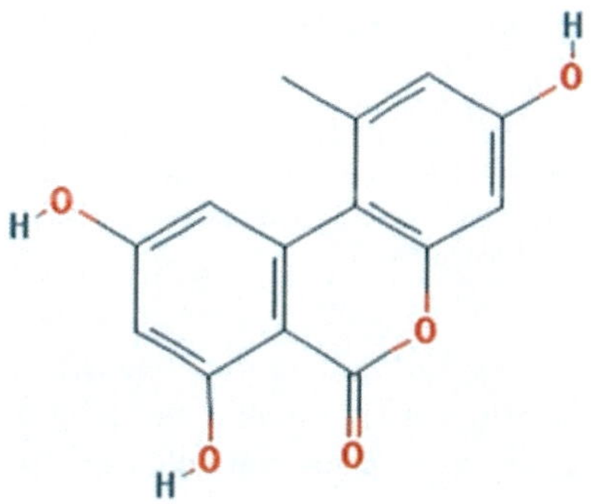

Fig. 3.17 Alternariol. (Adapted from PubChem)

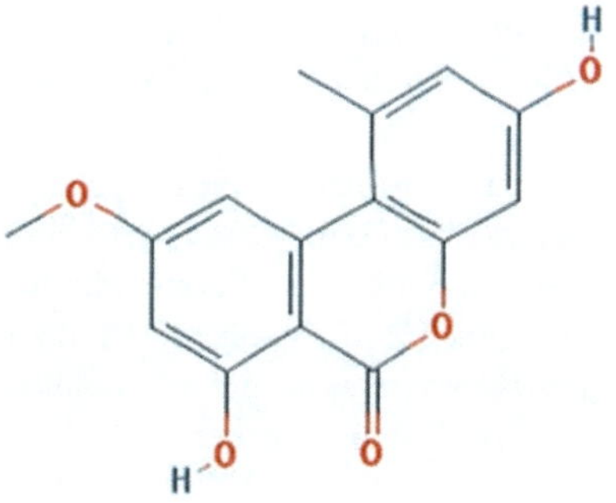

Fig. 3.18 AME. (Adapted from PubChem)

3.9.4 *Toxicity*

From a toxicological point of view, *Alternaria* toxins show cytotoxic and/or teratogenic activity, mutagenicity, clastogenicity, and oestrogenicity in either microbial or mammalian cell systems as well as tumorigenicity in rats and inhibition of cell proliferation.

Acute toxicity of AOH and AME are low in animals. Evidence of in vitro mutagenicity and carcinogenicity in exceptional assays has been recorded. When mice were fed with AME for 10 months, pre-cancerous changes in the oesophageal mucosa was observed. Squamous cell carcinoma in mice was induced when human embryo tissue was treated with AOH. AME transformed NIH/3T3 cells leading to the development of subcutaneous tumours in mice. In vitro AOH and AME induced breaks in DNA strands (Meena and Samal 2019; Kohmoto et al. 1976) (Table 3.9).

3.10 Regulation of Mycotoxins

The repercussions of mycotoxicoses are adverse and are long-lasting. They not only cause health issues but also result in huge economic losses. Mycotoxicosis not only hampers the image of the nation in the international market but simultaneously

Table 3.9 Toxicological profiles of AME and AOH

Toxicity	Mechanism of toxicological action
Cytotoxic, clastogenic, mutagenic, oestrogenic, and teratogenic in either microbial or mammalian cell systems and tumorigenic in rats (Meena and Samal 2019)	Inhibition of cell proliferation; AOH and AME cause breakage of DNA strands in cell cultures (Kohmoto et al. 1976)

erodes the health of the population. It leads to a ban on the export of products from developing countries, especially in the European Union region. There have been cases when European countries like the United Kingdom, France, and Italy have banned the export of mangoes, coffee, and many other fruits and vegetables from India during the past decade. The list also covers meat and fish products. Mycotoxicosis lowers animal productivity, leading to less production, which in turn curtails the volume of exports. The ban on exports has largely affected the export earnings, as well as the brand image of many developing countries in the international market, resulting in plummeting their position on the map of world exports. Furthermore, as the studies depict, reduction in levels of mycotoxins in food and feed will not only curtail the financial burden on health care but will also create new avenues for capturing the global market (Wu 2006).

Mycotoxin exposure is a serious risk to human health, particularly in developing countries where the effects of poverty and malnutrition overwhelm the harmful effects of these food-borne toxins by limiting biochemical reclamation mechanisms. Although the detrimental effects of mouldy grain to human health is known, the contaminated grain is often fed to livestock, leading to decreased animal productivity and the food supply and increased poverty (Turner et al. 2005). The exposure of mycotoxins should be considered a serious issue as they significantly restrict the attempts to improve human health in developing countries. The UN Millennium Development Goals directly aims at poverty issues in the developing world and hence on food choices and food safety (Gressel et al. 2004; Shephard 2008). Table 3.10 lists the regulatory limits of the major mycotoxins in the United States, European Union, Codex Alimentarius Commission, India, and China.

In communities in which food sufficiency cannot be attained, the consideration of mycotoxins as a major issue of food safety problem is difficult. In such a situation, it is unclear whether suitable technologies addressing precise subsistence farmers or the need of advanced technologies including improved varieties, genetically engineered organisms, and agricultural modelling or the blend of both the factors is the need of the hour to improve food safety (Gressel et al. 2004; Shephard 2008). Furthermore, the change in climatic conditions also affect the growth of fungi and hence the production of mycotoxins. Tropical conditions like high temperatures and moisture levels, unseasonal rains during harvests, monsoons, and flash floods enhance the fungal propagation, and the production of mycotoxins is enhanced in developing countries around the world. Fungal growth is prevalent in cereals grown in the humid subtropical climate regions of China, Thailand, Vietnam, and South Korea. Factors stimulating the production of mycotoxins are the humidity of the

Table 3.10 Permissible limit of mycotoxins in some countries (Jha 2015; Altech Mycotoxin Management; Mycotoxin Regulations in India (2011)—FSSAI, Mycotoxin-Regulations, http://www.knowmycotoxins.com/resources/mycotoxin-regulations)

Mycotoxin	United States	European Union	Codex Alimentarius Commission[a]	India	China[b]
Aflatoxin $B_1 + B_2 + G_1 + G_2$	Food: 20 ppb (corn, peanut products, cotton-seed meal, and other animal feed)	Food: 10 ppb [spices (*Capsicum* and *Piper* spp., nutmeg, ginger, and turmeric)]	15 ppb (peanuts intended for further processing)	30 ppb (all articles of food)[c]	Food: $AfB_1 \leq 20$ ppb (peanut and derived products; corn, cornmeal, and derived products; peanut and corn oils) $AFB_1 \leq 10$ ppb [rice, plant oils (except peanut and corn oils)] $AFB_1 \leq 5$ ppb (wheat and derived products; fermented soybean products; other cooked nuts; soy sauce, vinegar)
	Feed: 100 ppb (breeding beef cattle, breeding swine, or mature poultry)	Feed: 4 ppb (cereals and processed products thereof intended for direct consumption or as an ingredient)			Feed: $AfB_1 \leq 50$ ppb (corn products, peanut meal) $AfB_1 \leq 10$ ppb (plant oils except for peanut and corn oils) $AFB_1 \leq 20$ ppb (peanut and corn oils) $AFB_1 \leq 30$ ppb (other plant materials of feed)
Aflatoxin M_1	100 ppb (corn, peanut products, cottonseed meal, and other animal feed ingredients)	0.05 ppb (milk)	0.5 ppb (milk)	0.5 ppb (milk)	0.5 ppb (milk and dairy products)
Fumonisin $B_1+B_2+B_3$	20 ppb (food)				
	0.5 ppb (milk)				

(continued)

Table 3.10 (continued)

Mycotoxin	United States	European Union	Codex Alimentarius Commission[a]	India	China[b]
	300 ppb (finishing beef cattle)				$B_1+B_2 \leq 60$ ppm (feed materials such as corn and derived products, Dry Distillery Grain Soluble (DDGS), corn silage) $B_1+B_2 \leq 20$ ppm (complementary feedstuff for calf, lamb; poultry concentrated and complete feed) $B_1+B_2 \leq 5$ ppm (complementary feed for horse and rabbit; pig concentrated feed; complete feed for pig, rabbit, and horse)
DON	1 ppm (wheat products)	8 ppm (cereals and cereal products with the exception of maize by-products)			Food: 1 ppm (wheat and derived products; corn and cornmeal)
					Feed: 5 ppm (plant feed materials) 1 ppm (complementary feedstuff for calf, lamb, and lactating cows; complete feed for pig) 3 ppm (other complementary feedstuff and complete feed)
Ochratoxin A		3 ppb (cereal products)	20 ppb (wheat, barley, and rye)		Food: 5 ppb (cereal and derived products; bean and derived products; roasted coffee beans; ground coffee)

					2 ppb (grape wine) 10 ppb (instant coffee)
		5 ppb (raw cereal grains)			Feed: 100 ppb (cereal and derived products; formula feed)
		10 ppb [dried vine fruit (currants, raisins, sultanas)]			
Patulin	50 ppb (apple juice and its products)	50 ppb (apple juice and apple derived foods)	50 ppb (apple juice and apple juice ingredients in other beverages)	50 ppb (apple juice and apple juice ingredients in other beverages)	50 ppb (fruits and derived products; fruits and vegetable drink; and wine)
		2 ppb (groundnuts, nuts, dried fruits and processed products thereof intended for direct consumption)			Not applicable
Zearalenone					Food: 60 ppb (wheat, wheat flour, corn, cornmeal)
					Feed: 0.5 ppm (corn and derived products) 1.5 ppm (corn bran, corn starch, DDGS) 1.0 ppm (other plant feed materials)
Ergot alkaloids		1000 ppm (feed materials and compound feed containing unground cereals)			

[a] The Codex Alimentarius Commission (CAC) is an intergovernmental body founded by the Food and Agriculture Organization of the United Nations (FAO) and the World Health Organization (WHO) within the framework of the Joint Food Standards Programme having more than 180 members, to look after the health of consumers and ensure fair practices in the food trade

[b] China permissible limits data were kindly provided by Fuguo Xing, Chinese Academy of Agricultural Sciences

[c] Limits defined for aflatoxin B_1 only

substrate (10–20%), the relative humidity (≥70%), the temperature (0–50 °C, depending on the fungal species), and the availability of oxygen. The presence of high humidity and temperatures in tropical areas makes crops in tropical and subtropical areas more susceptible to mycotoxin contamination as compared to those in temperate zones.

3.11 Conclusion

Since the early 1960s, the impact of mycotoxigenic fungi on human pathogenesis has been the subject of exhaustive study. Most of the mycotoxins are known to hazardously contaminate crops and therefore animal feed and animal products leading to huge economic losses. Mycotoxins also have negative impacts on the health of animals and humans, the productivity of animals, and domestic and international trade. In humans, the presence of mycotoxins in foods is often cumulative and leads to cancers and immuno-deficiency diseases. Mycotoxin contamination can cause serious and sometimes fatal diseases. Most of the outbreaks of mycotoxicoses are caused by the ingestion of food contaminated with mycotoxins. Plant breeding, good agronomic practices, and detoxification can help in the prevention and control of mycotoxicosis. Therefore, strict control of food quality in industrialized and developing countries is required to avoid such outbreaks. Although the challenges of mycotoxin management seem difficult to overcome, a better understanding of our environment and adequate research will give us the skills to find the best solutions to attain a healthy mycotoxin-free life. In conclusion, it would be prudent for the medical community to pay closer attention to the possibility of mycotoxin toxicity in patients with symptoms of fungal exposure. Although the role of mycotoxins in human disease is still speculative, its acute mutagenicity, carcinogenicity, nephrotoxicity, teratogenic effects, and immunosuppressive actions in animal models, in combination with its ability to be carried through the food chain, is a matter of concern.

References

Abarca ML, Bragulat MR, Castella G, Cabanes FJ (1994) Ochratoxin A production by strains of *Aspergillu sniger* var. niger. Appl Environ Microbiol 60(7):2650–2652

Adhikari M, Negi B, Kaushik N, Adhikari A, Al-Khedhairy AA, Kaushik NK, Choi EH (2017) T-2 mycotoxin: toxicological effects and decontamination strategies. Oncotarget 8(20): 33933–33952. https://doi.org/10.18632/oncotarget.15422

Aflatoxin B1 (2003) http://www.fermentek.com/Aflatoxin_B1

Aflatoxin B2 (2003) http://www.fermentek.com/Aflatoxin B2

Aflatoxin G1 (2003) http://www.fermentek.com/product/aflatoxin-g1

Aflatoxin G2 (2003) http://www.fermentek.com/Aflatoxin G2

Aflatoxin M1 (2003) http://www.fermentek.com/product/aflatoxin-m1

Aflatoxin M2 (2006) http://www.fermentek.com/product/aflatoxin-m2

Akande KE, Abubakar MM, Adegbola TA, Bogoro SE (2006) Nutritional and health implications of mycotoxins in animal feeds: a review. Pak J Nutr 5(5):398–403

Alexander J, Benford D, Boobis A, Ceccatelli S, Cottrill B, Cravedi J, Di Domenico A, Doerge D, Dogliotti E, Edler L, Farmer P (2011) Scientific opinion on the risks for animal and public health related to the presence of *Alternaria* toxins in feed and food. Eur Food Safety Authority J 9(10): 2407–2504

Altomare C, Logrieco AF, Gallo A (2021) Mycotoxins and mycotoxigenic fungi: risk and management. A challenge for future global food safety and security. Encycl Mycol:64–93. https://doi.org/10.1016/B978-0-12-819990-9.00032-9

Alves I, Oliveira NG, Laires A, Rodrigues AS, Rueff J (2000) Induction of micronuclei and chromosomal aberrations by the mycotoxin patulin in mammalian cells: role of ascorbic acid as a modulator of patulin clastogenicity. Mutagenesis 15(3):229–234. https://doi.org/10.1093/mutage/15.3.229

Antonissen G, Martel A, Pasmans F, Ducatelle R, Verbrugghe E, Vandenbroucke V, Li S, Haesebrouck F, Van Immerseel F, Croubels S (2014) The impact of *Fusarium* mycotoxins on human and animal host susceptibility to infectious diseases. Toxins 6(2):430–452. https://doi.org/10.3390/toxins6020430

Bayman P, Baker JL, Doster MA, Michailides TJ, Mahoney NE (2002) Ochratoxin production by the *Aspergillus ochraceus* group and *Aspergillus alliaceus*. Appl Environ Microbiol 68(5): 2326–2329. https://doi.org/10.1128/AEM.68.5.2326-2329.2002

Bbosa GS, Kitya D, Odda J, Ogwal-Okeng J (2013) Aflatoxins metabolism, effects on epigenetic mechanisms and their role in arcinogenesis. Health 5(10):14. https://doi.org/10.4236/health.2013.510A1003

Beardall JM, Miller JD (1994) Disease in humans with mycotoxins as possible causes. In: Miller JD, Trenholm HL (eds) Mycotoxins in grains. Compounds other than aflatoxin. Eagan Press, St. Paul, Minn, pp 487–539

Becci PJ, Hess FG, Johnson WD, Gallo MA, Babish JG, Dailey RE, Parent RA (1981) Long-term carcinogenicity and toxicity studies of patulin in the rat. J Appl Toxicol 1(5):256–261. https://doi.org/10.1002/jat.2550010504

Bennett JW, Klich M (2003) Mycotoxins. Clin Microbiol Rev 16:497–516

Binder EM (2007) Managing the risk of mycotoxins in modern feed production. Anim Feed Sci Technol 133:149–166. https://doi.org/10.1016/j.anifeedsci.2006.08.008

Biomin (2021a) Mycotoxins. http://www.mycotoxins.info/en/mycotoxins/. Accessed 25 Sept 2021

Biomin (2021b) mycotoxins.info > Mycotoxins > Common Mycotoxins. http://www.mycotoxins.info/myco_info/science_moa.html. Accessed 19 July 2021

Bokhari F, Gherbawy Y, Najjar A (2009) Detection of the patulin-producing potential of some Paecilomyces variotii strains isolated from the air samples of Jeddah City, Saudi Arabia, using the RAPD-PCR technique. Aerobiologia 25(1):49–54. https://doi.org/10.1007/s10453-008-9108-0

Bourdiol D, Escoula L, Salvayre R (1990) Effect of patulin on microbicidal activity of mouse peritoneal macrophages. Food Chem Toxicol 28(1):29–33. https://doi.org/10.1016/0278-6915(90)90132-7

Boussabbeh M, Salem IB, Belguesmi F, Bacha H, Abid-Essefi S (2016) Tissue oxidative stress induced by patulin and protective effect of crocin. Neurotoxicology 53:343–349. https://doi.org/10.1016/j.neuro.2015.11.005

Bunge I, Heller K, Röschenthaler R (1979) Isolation and purification of ochratoxin A. Zeitschrift für Lebensmitteluntersuchung und-Forschung A 168(6):457–458

Carlson MP, Ensley SM (2003) Understanding fungal (mold) toxins (mycotoxins). Cooperative Extension, Institute of Agriculture and Natural Resources, University of Nebraska-Lincoln, Lincoln

Chalmers I, Clarke M (2004) Commentary: the 1944 patulin trial: the first properly controlled multicentre trial conducted under the aegis of the British Medical Research Council. Int J Epidemiol 33(2):253–260. https://doi.org/10.1093/ije/dyh162

Cheraghali AM, Mohammadi HR, Amirahmadi M, Yazdanpanah H, Abouhossain G, Zamanian F, Khansari MG, Afshar M (2005) Incidence of patulin contamination in apple juice produced in Iran. Food Control 16(2):165–167. https://doi.org/10.1016/j.foodcont.2004.01.006Get

Chu FS, Wilson BJ (1973) Studies on ochratoxins. CRC Crit Rev Toxicol 2(4):499–524. https://doi.org/10.3109/10408447309025706

Ciegler A, Fennell DI, Sansing GA, Detroy RW, Bennett GA (1973) Mycotoxin-producing strains of *Penicillium viridicatum*: classification into subgroups. Appl Microbiol 26(3):271–278

Dailey RE, Blaschka AM, Brouwer EA (1977) Absorption, distribution, and excretion of [14C] patulin by rats. J Toxicol Environ Health Part A Curr Issues 3(3):479–489. https://doi.org/10.1080/15287397709529580

Devaraj H, Suseela RE, Devaraj N (1986) Patulin toxicosis in chicks. Curr Sci 55(19):998–999

Döll S, Dänicke S (2011) The *Fusarium* toxins deoxynivalenol (DON) and zearalenone (ZON) in animal feeding. Prev Vet Med 102(2):132–145. https://doi.org/10.1016/j.prevetmed.2011.04.008

DON (2007) http://www.fermentek.com/Deoxynivalenol

El-Nezami H, Polychronaki N, Salminen S, Mykkänen H (2002) Binding rather than metabolism may explain the interaction of two food-grade *Lacto bacillus* strains with zearalenone and its derivative ά-zearalenol. Appl Environ Microbiol 68(7):3545–3549. https://doi.org/10.1128/AEM.68.7.3545-3549.2002

Ergotism mycotoxicoses, Merck veterinary manual. http://www.merckvetmanual.com/mvm/zk/toxicology/mycotoxicoses/ergotism.html. Accessed 8 July 2021

European Food Safety Authority (EFSA) (2005) Opinion of the Scientific Panel on Contaminants in Food Chain on a request from the Commission related to fumonisins as undesirable substances in animal feed. Eur Food Safety Authority J 235:1–32

European Food Safety Authority. Annual report 2011. https://www.efsa.europa.eu/en/corporate/pub/ar11. Accessed 15 April 2021

Fernández-Cruz ML, Mansilla ML, Tadeo JL (2010) Mycotoxins in fruits and their processed products: analysis, occurrence and health implications. J Adv Res 1(2):113–122. https://doi.org/10.1016/j.jare.2010.03.002

Food-Info.net> Topics > Food Safety > Toxins > Overview of food-borne toxins aflatoxins. http://www.food-info.net/uk/tox/afla.htm. Accessed 8 Sept 2021

Fumonisin B1 (2022) http://www.fermentek.com/product/fumonisin-b1

Gressel J, Hanafi A, Head G, Marasas W, Obilana AB, Ochanda J, Souissi T, Tzotzos G (2004) Major heretofore intractable biotic constraints to African food security that may be amenable to novel biotechnological solutions. Crop Prot 23(8):661–689. https://doi.org/10.1016/j.cropro.2003.11.014

Groopman JD, Kensler TW (2005) Role of metabolism and viruses in aflatoxin-induced liver cancer. Toxicol Appl Pharmacol 206(2):131–137. https://doi.org/10.1016/j.taap.2004.09.020

Haackgeb M (1978) On the relevance of ochratoxin A contamination in coffee from a toxicological point of view. Abschlussarbeit Postgradual stadium Toxikologie der Universität Leipzig. eingereicht von Dipl.-Biochemiker Michael Haackgeb. am 1. December 1978 in Halle/Saale eingereicht am 7. January 2008

Heidler D, Schatzmayr G (2003) A new approach to managing mycotoxins. Feed Mix 11(1):31–34

Hopmans EC (1997) Patulin: a mycotoxin in apples. Perishables Handling Q 91:5–6

Hussein HS, Brasel JM (2001) Toxicity, metabolism, and impact of mycotoxins on humans and animals. Toxicology 167(2):101–134. https://doi.org/10.1016/S0300-483X(01)00471-1

Ismaiel AA, Papenbrock J (2015) Mycotoxins: producing fungi and mechanisms of phytotoxicity. Agriculture 5(3):492–537. https://doi.org/10.3390/agriculture5030492

Jha SN (2015) Rapid detection of food adulterants and contaminants: theory and practice. Academic Press, Boston

Juan C, Oueslati S, Mañes J (2016) Evaluation of *Alternaria* mycotoxins in strawberries: quantification and storage condition. Food Addit Contam Part A 33(5):861–868. https://doi.org/10.1080/19440049.2016.1177375

Kenneth T (2006) The normal bacterial flora of humans. Todar's online textbook of bacteriology, pp 1–5. http://textbookofbacteriology.net/normalflora.html

Kharayat BS, Singh Y (2018) Mycotoxins in foods: mycotoxicoses, detection, and management. In: Microbial contamination and food degradation. Academic Press, Boston, pp 395–421. https://doi.org/10.1016/B978-0-12-811515-2.00013-5

Kohmoto K, Khan ID, Renbutsu Y, Taniguchi T, Nishimura S (1976) Multiple host-specific toxins of *Alternaria mali* and their effect on the permeability of host cells. Physiol Plant Pathol 8(2): 141–153. https://doi.org/10.1016/0048-4059(76)90047-3

Kőszegi T, Poór M (2016) Ochratoxin A: molecular interactions, mechanisms of toxicity and prevention at the molecular level. Toxins 8(4):111. https://doi.org/10.3390/toxins8040111

Kuiper-Goodman T, Scott PM (1989) Risk assessment of the mycotoxin ochratoxin A. Biomed Environ Sci 2(3):179–248

Kuiper-Goodman T, Hilts C, Billiard SM, Kiparissis Y, Richard IDK, Hayward S (2010) Health risk assessment of ochratoxin A for all age-sex strata in a market economy. Food Addit Contam 27(2):212–240

Liu BH, Wu TS, Yu FY, Su CC (2007) Induction of oxidative stress response by the mycotoxin patulin in mammalian cells. Toxicol Sci 95(2):340–347. https://doi.org/10.1093/toxsci/kfl156

Liu BH, Yu FY, Wu TS, Li SY, Su MC, Wang MC, Shih SM (2003) Evaluation of genotoxic risk and oxidative DNA damage in mammalian cells exposed to mycotoxins, patulin and citrinin. Toxicol Appl Pharmacol 191(3):255–263. https://doi.org/10.1016/S0041-008X(03)00254-0

Llewellyn GC, McCay JA, Brown RD, Musgrove DL, Butterworth LF, Munson AE, White KL Jr (1998) Immunological evaluation of the mycotoxin patulin in female B6C3F1 mice. Food Chem Toxicol 36(12):1107–1115. https://doi.org/10.1016/S0278-6915(98)00084-2

Mandappa IM, Basavaraj K, Manonmani HK (2018) Analysis of mycotoxins in fruit juices. In: Fruit juices. Academic Press, Boston, pp 763–777. https://doi.org/10.1016/B978-0-12-802230-6.00036-9

Marin S, Ramos AJ, Cano-Sancho G, Sanchis V (2013) Mycotoxins: occurrence, toxicology, and exposure assessment. Food Chem Toxicol 60:218–237. https://doi.org/10.1016/j.fct.2013.07.047

Marquardt RR, Frohlich AA (1992) A review of recent advances in understanding ochratoxicosis. J Anim Sci 70(12):3968–3988. https://doi.org/10.2527/1992.70123968x

McKinley ER, Carlton WW (1980a) Patulin mycotoxicosis in Swiss ICR mice. Food Cosmet Toxicol 18(2):181–187. https://doi.org/10.1016/0015-6264(80)90072-3

McKinley ER, Carlton WW (1980b) Patulin mycotoxicosis in the Syrian hamster. Food Cosmet Toxicol 18(2):173–179. https://doi.org/10.1016/0015-6264(80)90071-1

McKinley ER, Carlton WW, Boon GD (1982) Patulin mycotoxicosis in the rat: toxicology, pathology and clinical pathology. Food Chem Toxicol 20(3):289–300. https://doi.org/10.1016/S0278-6915(82)80295-0

Meena M, Samal S (2019) Alternaria host-specific (HSTs) toxins: an overview of chemical characterization, target sites, regulation and their toxic effects. Toxicol Rep 6:745–758. https://doi.org/10.1016/j.toxrep.2019.06.021

Murillo-Arbizu M, Amézqueta S, González-Peñas E, de Cerain AL (2009) Occurrence of patulin and its dietary intake through apple juice consumption by the Spanish population. Food Chem 113(2):420–423. https://doi.org/10.1016/j.foodchem.2008.07.054

Muro-Cach CA, Stedeford T, Banasik M, Suchecki TT, Persad AS (2004) Mycotoxins: mechanisms of toxicity and methods of detection for identifying exposed individuals. J Land Use Environ Law 19(2):537–556

Mycotoxin Regulations in India, FSSAI (2011) https://foodregulatory.fssai.gov.in/All%20Docs/Food%20Standards/regulations/Contaminants_Regulations.pdf

Mycotoxin-Regulations. http://www.knowmycotoxins.com/resources/mycotoxin-regulations

Mycotoxins. Power point. site.iugaza.edu.ps/akichaoi/files/2010/02/Mycotoxin.ppt.ppt. Accessed 13 July 2021

Nair MG (1998) Fumonisins and human health. Ann Trop Paediatr 18(suppl 1):S47–S52. https://doi.org/10.1080/02724936.1998.11747980

National Center for Biotechnology Information (2022a) PubChem Compound Summary for CID 5359485, Alternariol. https://pubchem.ncbi.nlm.nih.gov/compound/Alternariol. Accessed 9 May 2022

National Center for Biotechnology Information (2022b) PubChem Compound Summary for CID 8223, Ergotamine. https://pubchem.ncbi.nlm.nih.gov/compound/Ergotamine. Accessed 9 May 2022

National Center for Biotechnology Information (2022c) PubChem Compound Summary for CID 442530, Ochratoxin A. https://pubchem.ncbi.nlm.nih.gov/compound/Ochratoxin-A. Accessed 9 May 2022

Ochratoxin A (2020) http://www.fermentek.com/product/ochratoxin-benzene-free

Ostry V (2008) *Alternaria* mycotoxins: an overview of chemical characterization, producers, toxicity, analysis and occurrence in foodstuffs. World Mycotoxin J 1(2):175–188. https://doi.org/10.3920/WMJ2008.x013

Pal S, Singh N, Ansari KM (2017) Toxicological effects of patulin mycotoxin on the mammalian system: an overview. Toxicol Res 6(6):764–771. https://doi.org/10.1039/c7tx00138j

Papatsiros VG (2012) The splay leg syndrome in piglets: a review. Am J Vet Sci 7:80–83

Park DL (1993) Controlling aflatoxin in food and feed. Food Technol 47:92

Patulin (2021) http://www.chemspider.com/Chemical-Structure.4534.html. Accessed 15 Oct 2021

Pitt JI (1987) *Penicillium viridicatum, Penicillium verrucosum*, production of ochratoxin A. Appl Environ Microbiol 53(2):266–269

Pitt JI (2000) Toxigenic fungi: which are important? Med Mycol 38(suppl 1):17–22

Placinta CM, D'mello JPF, Macdonald AMC (1999) A review of worldwide contamination of cereal grains and animal feed with *Fusarium* mycotoxins. Anim Feed Sci Technol 78(1):21–37. https://doi.org/10.1016/S0377-8401(98)00278-8

Pose G, Patriarca A, Kyanko V, Pardo A, Pinto VF (2010) Water activity and temperature effects on mycotoxin production by *Alternaria alternata* on a synthetic tomato medium. Int J Food Microbiol 142(3):348–353. https://doi.org/10.1016/j.ijfoodmicro.2010.07.017

Puel O, Galtier P, Oswald IP (2010) Biosynthesis and toxicological effects of patulin. Toxins 2(4): 613–631. https://doi.org/10.3390/toxins2040613

Ramalingam S, Bahuguna A, Kim M (2019) The effects of mycotoxin patulin on cells and cellular components. Trends Food Sci Technol 83:99–113. https://doi.org/10.1016/j.tifs.2018.10.010

Reddy CS, Chan PK, Hayes AW (1978) Teratogenic and dominant lethal studies of patulin in mice. Toxicology 11:219–223. https://doi.org/10.1016/S0300-483X(78)91339-2

Reddy SV, Waliyar F (2000) Properties of aflatoxin and it producing fungi. Aflatoxin. International Crops Research Institute for the Semi-Arid Tropics

Rheeder JP, Marasas WF, Vismer HF (2002) Production of fumonisin analogs by *Fusarium* species. Appl Environ Microbiol 68(5):2101–2105. https://doi.org/10.1128/AEM.68.5.2101-2105.2002

Richard JL (2007) Some major mycotoxins and their mycotoxicoses-an overview. Int J Food Microbiol 119(1):3–10. https://doi.org/10.1016/j.ijfoodmicro.2007.07.019

Riley RT, Showker JL (1991) The mechanism of patulin's cytotoxicity and the antioxidant activity of indoletetramic acids. Toxicol Appl Pharmacol 109(1):108–126. https://doi.org/10.1016/0041-008X(91)90195-K

Roll R, Matthiaschk G, Korte A (1990) Embryotoxicity and mutagenicity of mycotoxins. J Environ Pathol Toxicol Oncol 10(1-2):1–7

Ryu D, Hanna MA, Bullerman LB (1999) Stability of zearalenone during extrusion of corn grits. J Food Prot 62(12):1482–1484. https://doi.org/10.4315/0362-028X-62.12.1482

Saleh I, Goktepe I (2019) The characteristics, occurrence, and toxicological effects of patulin. Food Chem Toxicol 129:301–311. https://doi.org/10.1016/j.fct.2019.04.036

Samapundo S, Devlieghere F, De Meulenaer B, Geeraerd AH, Van Impe JF, Debevere JM (2005) Predictive modelling of the individual and combined effect of water activity and temperature on the radial growth of *Fusarium verticillioides* and *F. proliferatum* on corn. Int J Food Microbiol 105:35–52. https://doi.org/10.1016/j.ijfoodmicro.2005.06.007

Sanderson PG, Spotts RA (1995) Postharvest decay of winter pear and apple fruit caused by species of *Penicillium*. Phytopathology 85(1):103–110

Schumacher DM, Wagner J, Metzler M, Lehmann L (2005) Influence of decreased intracellular glutathione level on the mutagenicity of patulin in cultured mouse lymphoma cells. Mycotoxin Res 21(2):150–152

Scott PM, Stoltz DR (1980) Mutagens produced by *Alternaria alternata*. Mutat Res 78(1):33–40. https://doi.org/10.1016/0165-1218(80)90023-3

Shephard GS (2008) Impact of mycotoxins on human health in developing countries. Food Addit Contam 25(2):146–151. https://doi.org/10.1080/02652030701567442

Sigma-Aldrich webpage (2021) Patulin http://www.sigmaaldrich.com/catalog/product/sigma/p1639?lang=en&region=US

Smith EE, Duffus EA, Small MH (1993) Effects of patulin on postimplantation rat embryos. Arch Environ Contam Toxicol 25(2):267–270

T2 toxin (2004) http://www.fermentek.com/product/t2-toxin

Tournas VH, Katsoudas E (2005) Mould and yeast flora in fresh berries, grapes and citrus fruits. Int J Food Microbiol 105(1):11–17. https://doi.org/10.1016/j.ijfoodmicro.2005.05.002

Turner PC, Sylla A, Gong YY, Diallo MS, Sutcliffe AE, Hall AJ, Wild CP (2005) Reduction in exposure to carcinogenic aflatoxins by postharvest intervention measures in west Africa: a community-based intervention study. Lancet 365(9475):1950–1956. https://doi.org/10.1016/S0140-6736(05)66661-5

Umeda M, Tsutsui T, Saito M (1977) Mutagenicity and inducibility of DNA single-strand breaks and chromosome aberrations by various mycotoxins. Gann Jpn J Cancer Res 68(5):619–625. https://doi.org/10.20772/cancersci1959.68.5_619

Urry WH, Wehrmeister HL, Hodge EB, Hidy PH (1966) The structure of zearalenone. Tetrahedron Lett 7(27):3109–3114. https://doi.org/10.1016/S0040-4039(01)99923-X

Van Egmond HP, Speijers GJA (1994) Survey of data on the incidence and levels of ochratoxin A in food and animal feed worldwide. J Nat Toxins 3(2):125–144

Vidal A, Ouhibi S, Ghali R, Hedhili A, De Saeger S, De Boevre M (2019) The mycotoxin patulin: an updated short review on occurrence, toxicity and analytical challenges. Food Chem Toxicol 129:249–256. https://doi.org/10.1016/j.fct.2019.04.048

Wannemacher RW, Wiener SL, Sidell FR, Takafuji ET, Franz DR (1997) Trichothecene mycotoxins. Med Aspects Chem Biol Warfare 6:655–676

WHO (1998) Joint FAO/WHO Expert Committee on Food additives (JECFA), Position paper on patulin, 30th session, The Hague, The Netherlands, 9–13 March 1998

Wild CP, Gong YY (2009) Mycotoxins and human disease: a largely ignored global health issue. Carcinogenesis 31(1):71–82. https://doi.org/10.1093/carcin/bgp264

Wilson SC, Brasel TL, Martin JM, Wu C, Andriychuk L, Douglas DR, Cobos L, Straus DC (2003) Efficacy of chlorine dioxide as a gas and in solution in the inactivation of two trichothecene mycotoxins. Int J Toxicol 24(3):181–186

World Health Organization, International Agency for Research on Cancer (1993) Some naturally occurring substances: food items and constituents, heterocyclic aromatic amines and mycotoxins. IARC Monographs on the evaluation of the carcinogenic risk of chemicals to humans, p 56

Wu F (2006) Mycotoxin reduction in Bt corn: potential economic, health, and regulatory impacts. ISB News Report, Sept 2006. https://doi.org/10.1007/s11248-005-5237-1

Wu TS, Yang JJ, Yu FY, Liu BH (2012) Evaluation of nephrotoxic effects of mycotoxins, citrinin and patulin, on zebrafish (*Danio rerio*) embryos. Food Chem Toxicol 50(12):4398–4404. https://doi.org/10.1016/j.fct.2012.07.040

Wu L, Qiu L, Zhang H, Sun J, Hu X, Wang B (2017) Optimization for the production of deoxynivalenol and zearalenone by *Fusarium graminearum* using response surface methodology. Toxins 9(2):57. https://doi.org/10.3390/toxins9020057

Yuan Y, Zhuang H, Zhang T, Liu J (2010) Patulin content in apple products marketed in Northeast China. Food Control 21:1488–1491. https://doi.org/10.1016/j.foodcont.2010.04.019

Zain ME (2011) Impact of mycotoxins on humans and animals. J Saudi Chem Soc 15(2):129–144. https://doi.org/10.1016/j.jscs.2010.06.006

Zearelenone (2002) http://www.fermentek.com/product/zearalenone

Chapter 4
Emerging Mycotoxins and Their Clinicopathological Effects

The term "emerging mycotoxins" was coined in 2008 for lesser-known *Fusarium* metabolites, fusaproliferin, enniatins, beauvericin, and moniliformin by Jestoi (2008). The emerging mycotoxins and their derivatives can occur in high concentrations and high frequency in cereals, cereal-based products, fruits, processed and unprocessed/raw food, animal feed, and feeding stuffs. Some of them are carried over in animal-derived products. Notwithstanding the high prevalence, determination and legislative regulations of emerging mycotoxins are confined to Europe and the Mediterranean (Santini et al. 2012). Since many emerging mycotoxins are non or less toxic to the consumers, their importance to food safety is not addressed properly. In view of this, the present chapter focuses on some emerging or minor mycotoxins produced by *Aspergillus* spp., *Penicillium* spp., *Fusarium* spp., and *Alternaria* spp.

4.1 Beauvericin

Beauvericin (BEA) is produced by *Fusarium proliferatum*, *F. oxysporum*, *F. subglutinans*, *F. verticilliodes*, *F. circinatum* (in the USA and South Africa), *F. sambucinum* (Europe), *F. bulbieola*, *F. lactis*, *F. pseudicircinatum*, *F. denticulatum*, *F. succisae*, *F. avenaceum*, *F. phyllophillum*, and an entomopathogenic fungus *Beauveria bassiana* (Logrieco et al. 2002; Santini et al. 2012). It is an emerging mycotoxin and has been reported to contaminate food products and feed stuff across the globe.

4.1.1 Chemical Structure

Chemically, BEA is a cyclic hexadepsipeptide ((3S,6R,9S,12R,15S,18R)-3,9,15-tribenzyl-4,10,16-trimethyl-6,12,18-tri(propan-2-yl)-1,7,13-trioxa-4,10,16-

K. Singh, A. Kumari, *Mycotoxins and Mycotoxicoses*,
https://doi.org/10.1007/978-981-19-2370-8_4

triazacyclooctadecane-2,5,8,11,14,17-hexone) (Fig. 4.1). It is structurally similar to ennatin and belongs to the ENN family. BEA is an ionophore and is composed of methylphenylalanyl and hydroxyvaleryl residues (EMBL-EBI n.d.). The molecular formula of beauvericin is $C_{45}H_{57}N_3O_9$. Its molecular weight is 783.9 (PubChem n.d.-a). This emerging mycotoxin has three phenylalanine residues (Firáková et al. 2007). The nucleophilic nature of BEA helps in the formation of dimeric structures thus increasing cell permeability (Luz et al. 2017).

4.1.2 Occurrence

Beauvericin is a natural contaminant of cereal-based products, compound feeds, supplement forages, and beans from organic farming (Juan et al. 2013; Křížová et al. 2021). The high incidence of BEA in wheat and wheat-based products like baby food, biscuits, pasts, breakfast cereals, pizza dough (refrigerated), rice, barley, oats, and rye has been reported from several countries (Stanciu et al. 2017; Decleer et al. 2018). Beauvericin was also found in the Swiss waste water treatment plant (Schenzel et al. 2012) and in the four Danish cereals viz. wheat, barley, oats, and rye (Svingen et al. 2017). The occurrence of BEA has also been detected in rice and rice-based baby food and maize from Morocco (Mahnine et al. 2011), wheat-based products from Romania (Stanciu et al. 2017), in commercial pet food and medicinal herbs from China (Hu and Rychlik 2014), and in peanuts and peanut paste from France (Manizan et al. 2018). The carry-over of BEA has been detected in the egg yolk, some tissues of turkeys, skin, thigh muscles and liver of broilers and breast milk in humans (Callebaut et al. 2011; Braun et al. 2020).

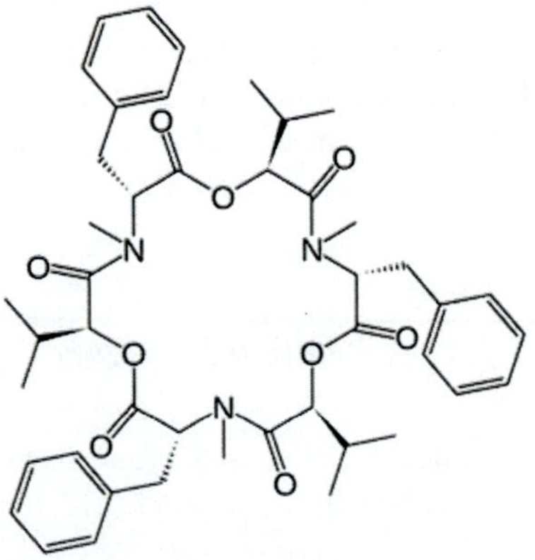

Fig. 4.1 Chemical structure of beauvericin. (Adapted from Fvasconcellos 22:20, 18 November 2007 (UTC)—Own work, Public Domain, https://commons.wikimedia.org/w/index.php?curid=3115419)

4.1.3 *Toxicity*

BEA is a phytotoxic compound. It is cytotoxic and can induce apoptosis by increasing ROS generation. Beauvericin can cross BBB (blood-brain barrier), thus causing neurotoxicity (Taevernier et al. 2016). It has an antagonistic effect on androgen receptors and can elevate the level of intracellular calcium (García-Herranz et al. 2019). BEA has been found immunotoxic to female mice (Maranghi et al. 2018) and exhibited high hepatotoxic potential in a quadroprobe hepatotoxicity assay (Svingen et al. 2017).

Nevertheless, beauvericin has anti-tumour, anti-bacterial, anti-fungal, insecticidal, and anti-viral properties, but at the same time, it is an existential threat. As there is limited data available on BEA, therefore, its risk assessment is still to be identified.

4.2 Butenolide

Butenolide is one of the *Fusarium* mycotoxins, which is produced by *Fusarium tricinctium*, *F. graminearum*, and *F. equiseti*. The natural co-occurrence of BUT and DON in cereals and cereal-based products has been reported globally (Woelflingseder et al. 2020).

4.2.1 *Chemical Structure*

Butenolide (BUT) is 4-acetamido-4-hydroxy-2-butenoic acid gamma lactone (Blackwell et al. 1985). The molecular formula of BUT is $C_6H_7O_3$ (Fig. 4.2). Its IUPAC name is 2-[(2R)-2-methyl-5-oxo-2H-furan-4-yl] acetic acid. The molecular weight of butenolide is 156.14 (PubChem n.d.-c).

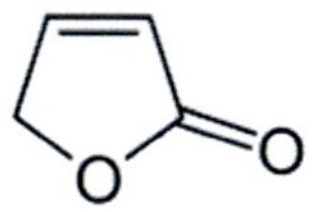

Fig. 4.2 Chemical structure of butenolide. (By Xplus1—Own work, Public Domain, https://commons.wikimedia.org/w/index.php?curid=47656322)

4.2.2 Occurrence

BUT has been detected in the corn, fescue, feed, and feed raw materials (Streit et al. 2013).

4.2.3 Toxicity

BUT has been shown to impair the lipid peroxidation and antioxidant system of myocardium in the chick embryo (Wang et al. 2008a, b) and also induce oxidative damage in the liver and kidneys (Wang et al. 2009a). Butenolide has possibly been related to Keshan disease which is prevalent in the remote mountains of China (Shi et al. 2021). The clinical symptoms of the disease are heart failure (congestive and acute) and cardiac arrhythmia. BUT has been reported to induce cytotoxicity in HepG2 cell line (Woelflingseder et al. 2020), cardiomyocytes of mouse and rat neonates (Yang et al. 2010) and cardiomyocyte-mitochondrial dysfunction in rats (Wang et al. 2009b). Similarly, 10–20 mg/kg of BUT can cause liver and myocardium injuries and weight loss in rats (Wang et al. 2009a, b, c). The application of butenolide ($\geq$1.5 mg) on the skin of heifers caused localized encrusted inflammatory lesions (Kosuri et al. 1970) while 4.5 g/day *per os* for 2 months and 3.8 mg/day IM for 3 months of BUT caused gastric and oesophageal ulcers, weight loss and tail-tip necrosis in cows. However, the oral administration of butenolide (at 39 and 68 mg) in young male calves (steers) was found lethal (Kosuri et al. 1970; Tookey et al. 1972). Although experimental evidence of BUT-toxicity has been reported, a thorough assessment of its toxicological profile is still wanted.

4.3 Citrinin

Citrinin (CIT) was first isolated from *Penicillium citrinum* Thom, in 1931 (Hetherington and Raistrick 1931; Gore et al. 1946). Later on, the other species of *Penicillium viz. P. citrinum*, *P. expansum*, *P. radicicola*, *P. camemberti*, *P. verrucosum*, *P. citreoviridae*, *P. atrium*, and *P. islandicum* and *Aspergillus like A. terreus*, *A. oryzae*, and *A. niveus* were also found to produce CIT. It has also been isolated from *Monascus purpureus* and *M. ruber* (Bennett and Klich 2003; Ostry et al. 2013; Silva et al. 2020). In the beginning, CIT was named monascidin A (Gore et al. 1946).

4.3.1 Chemical Structure

CIT is an acidic polyketide-derived mycotoxin. Its molecular weight is 250.25. The chemical formula of CIT is $C_{13}H_{14}O_5$ (Fig. 4.3). The IUPAC name of this lemon yellow crystalline compound is (3R,4S)-6-hydroxy-3,4,5-trimethyl-8-oxo-3,4-dihydroisochromene-7-carboxylic acid (PubChem n.d.-d). Citrinin decomposes into comparatively more cytotoxic citrinin H1 and less cytotoxic citrinin H2 (Hirota et al. 2002).

4.3.2 Occurrence

Citrinin has been detected in cereals and cereal-based products, oat, rye, barley, corn, and wheat grains, food supplements based on fermented rice (with *M. purpureus*), beans, black olives, apples, pomaceous fruits, oranges, tomatoes, sweet cherries, fruit and vegetable juices, roasted nuts, herbs, spices, edible tissues and eggs of poultry, animal-derived products like cheese, spoiled dairy products, beer, ham, sufu as well as fermented meat products (Li et al. 2012; Ostry et al. 2013; Markov et al. 2013; Lhotská et al. 2016; Wang et al. 2016; Dalefield 2017; Huiqin et al. 2017; Silva et al. 2020).

4.3.3 Toxicity

CIT predominantly causes nephrotoxicity. However, it also shows fetotoxicity in mammals (Flajs and Peraica 2009). Citrinin-induced chronic toxicity exhibited non-specific symptoms like weight loss, dehydration, polyuria, and polydipsia. CIT rarely causes acute toxicity but can affect the liver, heart, and reproductive system in mammals (de Oliveira Filho et al. 2017). The dietary exposure of citrinin caused watery faeces in poultry. Apart from this, the histopathological changes (degeneration and necrosis) have been seen in the liver and kidneys (Dalefield 2017). It has also been reported to induce nephropathy in pigs and other animal species. The co-contamination of CIT and OTA has resulted in the development of the C-C8dG-OTA DNA adduct (modified DNA adduct) (Ostry et al. 2013). Citrinin exerts less severe genotoxic effects than other *Penicillium* mycotoxin, the patulin

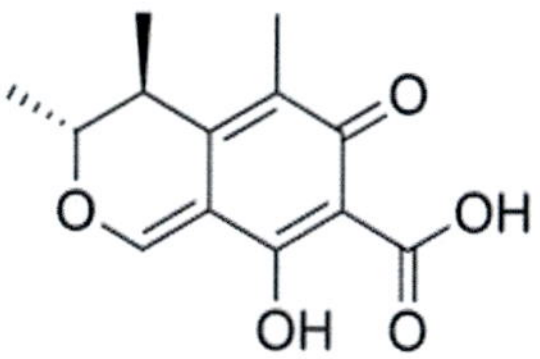

Fig. 4.3 Chemical structure of citrinin. (By Ed (Edgar181)—Own work, Public Domain, https://commons.wikimedia.org/w/index.php?curid=14732445)

(Iwahashi et al. 2007). The LD_{50} of CIT for ducks, chickens, and rabbits are 57 mg/kg, 95 mg/kg, and 134 mg/kg respectively (Bennett and Klich 2003).

The EU, Japan, and China have a permissible limit of 50 μg/kg, 200 μg/kg, and 2000 μg/kg respectively for citrinin in rice-based food supplements. But its maximum limit for other food products and feeding stuff has not been established yet (Silva et al. 2020). As citrinin is a potential nephrotoxin and can cause primary as well as secondary mycotoxicoses in humans, therefore, its levels should be regulated for proper risk assessment (Gil-Serna et al. 2014a).

4.4 Culmorin

Culmorin (CUL) is produced by *Fusarium venenatum*, *F. culmorum*, *F. crookwellense*, *F. graminearum*, *F. venenatum*, *F. praegraminearum*, and *Leptosphaeria oraemaris*, a marine ascomycete (Weber et al. 2018; Pedersen and Miller 1999; Strongman et al. 1987). CUL was first time isolated in the year 1937 but its structure could be elucidated in 1967 (Ashley et al. 1937; Barton and Werstiuk 1967). Several derivatives of CUL like 5-, 12-, 15-hydroxyculmorin, hydroxylculmorones, and culmorones have also been identified (Langseth et al. 2001).

4.4.1 *Chemical Structure*

Culmorin is a mycotoxin with the chemical formula $C_{15}H_{26}O_2$. The molecular weight of CUL is 238.37 Its IUPAC name is 2,6,6,9-tetramethyltricyclo [5.4.0.02,9]undecane-8,11-diol (Fig. 4.4) (PubChem n.d.-e).

4.4.2 *Occurrence*

The natural occurrence of culmorins with trichothecenes has been reported in *Fusarium* contaminated grains (Pedersen and Miller 1999), wheat, barley, oats,

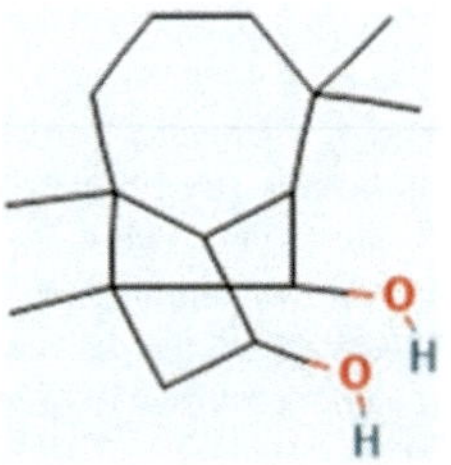

Fig. 4.4 Chemical structure of culmorin. (PubChem n.d.-e)

and mixed feed (Ghebremeskel and Langseth 2001). The derivatives of CUL have frequently been detected in malting barley, cereals, ground, and other nuts and their processed products (Abia et al. 2013).

4.4.3 *Toxicity*

Culmorin is less toxic than other *Fusarium* mycotoxins. It has phytotoxicity. Limited data is available on the toxicity of CUL in animals. Its high concentration, however, showed mild toxicity on BHK-21 kidney cells of baby hamster (Miller and MacKenzie 2000). The LD_{50} values of culmorin for chick embryo were reported to be 68 μg/kg and 78.2 μg/egg (Prelusky et al. 1989; Miller and MacKenzie 2000).

Although the toxicity of CUL has not been reported in animals and humans its frequent co-occurrence with DON in cereals should be comprehensively investigated. Moreover, the in vivo studies should be conducted to elucidate the complete toxicological profile of CUL.

4.5 Cyclopiazonic Acid

Cyclopiazonic acid (CPA) is a neurotoxic mycotoxin known to be produced by several species of *Aspergillus* and *Penicillium* (Chang et al. 2009). CPA was originally isolated from *Penicillium cyclopium* and then from a number fungi including *Penicillium griseofulvum*, *Penicillium camemberti*, *Penicillium commune*, *Aspergillus flavus*, *Aspergillus versicolor*, and *Aspergillus tamarii* (Chang et al. 2009; Krska et al. 2012).

CPA has been found to co-occur with ochratoxin A, patulin, citrinin, and zearalenone (Keblys et al. 2004; Krska et al. 2012). The highest production of CPA takes place at a temperature of 25 ± 1 °C (Gqaleni et al. 1997; Ostry et al. 2018). Presence of humidity >14% and condensation in the storage facility promotes mould growth and hence the production of CPA (Krska et al. 2012). The conditions favourable for the formation of CPA are similar to those of aflatoxins. CPA (also known as α-CPA) is biosynthesized using three enzymes; the polypeptide CpaS, dimethylallyl transferase (CpaD), and flavoprotein oxidocyclase (CpaO) (Liu and Walsh 2009).

4.5.1 *Chemical Structure*

Cyclopiazonic acid (Fig. 4.5) is a heavily substituted indole tetrameric acid (Chang et al. 2009). Its empirical formula is $C_{20}H_{20}N_2O_3$. The molecular weight is 336.391 g mol^{-1} and IUPAC nomenclature is (6a*R*,11a*S*,11b*R*)-10-Acetyl-11-

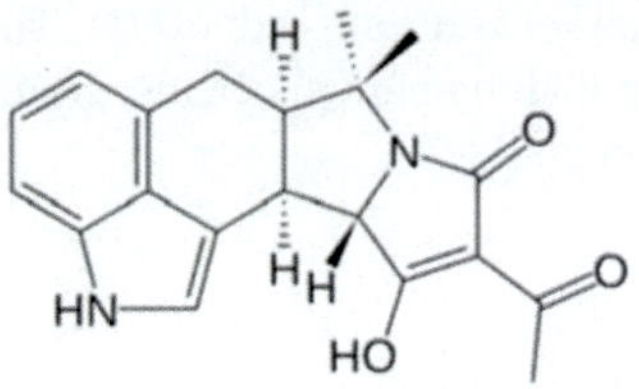

Fig. 4.5 Chemical structure of cyclopiazonic acid. (By Fvasconcellos 18:08, 26 September 2007 (UTC)—Own work, Public Domain, https://commons.wikimedia.org/w/index.php?curid=2818093)

hydroxy-7,7-dimethyl-2,6,6a,7,11a,11b-hexahydro-9*H*-pyrrolo[1′,2′:2,3]isoindolo [4,5,6-*cd*]indol-9-one (PubChem n.d.-b). It is soluble in dimethyl sulfoxide (DMSO) and ethanol but is insoluble in water (Keblys et al. 2004). CPA can chelate various cations such as calcium, magnesium, and iron, which might be an important mechanism of its toxicity (Keblys et al. 2004; Bryden 2009).

4.5.2 *Occurrence*

CPA is widely distributed in the environment and has been detected in naturally contaminated agricultural raw materials and mixed animal feeds (Sláma et al. 2008). CPA producing microfungi are able to grow different agricultural products, including grains, corn, sunflower seeds, peanuts, and cheese (Sláma et al. 2008). A natural occurrence of CPA has been noted in corn and a type of millet (kodo). The residues of CPA have been found in commercial milk, sheep milk, and in eggs and tissues of poultry (Dorner et al. 1994; Oliveira et al. 2006). The dietary exposure of cyclopiazonic acid in cattle and sheep can be carried over in their meat (Dorner et al. 1994).

4.5.3 *Toxicity*

Effects of CPA might often be masked by aflatoxicosis (Bullerman 2003). CPA is not a potent acute toxin as very high concentrations of CPA are required to cause mycotoxicosis in animals. CPA is toxic to both humans and animals when dietary exposure occurs through food and feed. Severe gastrointestinal distress and neurological disorders have been documented in the chicken, dog, rabbit, rat, and pig after consumption of CPA-contaminated feed. The degenerative changes and necrosis were also observed in the digestive tract, liver, kidney, and heart of these animals (Bhatnagar et al. 2014). The skeletal muscle degeneration (myofibrillar swelling and fragmentation) was observed in the broiler chickens fed with CPA. Cyclopiazonic acid-induced toxicity has been reported to cause hypokinesia, hypothermia, catalepsy, tremors, irregular respiration, ptosis, weight loss, and eventual death in mice (Bianchini and Bullerman 2014). CPA can weaken the calcium metabolism by inhibiting Ca^{2+}-activated ATPase activity and causing neural damage (Nishie et al. 1985; Goeger and Riley 1989). The symptomatology of CPA intoxication in humans includes anorexia, diarrhoea, pyrexia, dehydration, weight loss, ataxia, immobility, and extensor spasm at the time of death. Moreover, the histopathological manifestations such as alimentary tract hyperaemia, haemorrhage, and focal ulceration along with focal necrosis in the liver, pancreas, spleen, kidneys, and heart muscles were also observed (Gil-Serna et al. 2014a).

The possible association of CPA with turkey "X" and kodo poisoning has led to certain regulations by the European Food Safety Authority in 2009 (EFSA 2011).

4.6 Emodin

Emodin (EMO) is a crystalline compound and orange-red in colour. It is produced by *Penicillium rergulosum*, *Cladosporium fulvum*, *Aspergillus wentii*, and *Aspergillus ochraceous* (Agosti et al. 1962; Birkenshaw 1955; Hasan 1998). Emodin is a precursor of naturally occurring emodin-8-glucoside and other anthraquinones such as rhein, chrysophanol, and aloe-emodin (Wells et al. 1975). Emodin was isolated for the first time in 1925 as a metabolite of *Dermocybe sanguine*, "fragula-emodin" (Kögl and Postowsky 1925). Since then, emodin has been isolated as pigmented metabolites of *Penicilliopsis clavariae-forms*, *Cladosporium fulvum*, *Penicillium avellaneum*, *Aspergillus ochraceous*, and *Talaromyces avellanens* (Frisvad and Larsen 2015; Agosti et al. 1962; Natori et al. 1965).

4.6.1 Chemical Structure

Emodin is 1,3,8-tryhydroxy-6-methyl anthraquinone (Fig. 4.6). The chemical formula of emodin is $C_{15}H_{10}O_5$. Its molecular weight is 270.24 g/mol. EMO is soluble in alcohol, ammonia and sodium carbonate while moderately soluble in aqueous alkali hydroxide solutions. Emodin is insoluble in water (PubChem n.d.-f). Its melting point is 256–257 °C. The two metabolites of EMO, 2 hydroxyemodin and omega-hydroxy emodin, were reported to form in liver microsomes of rat (Mueller et al. 1996).

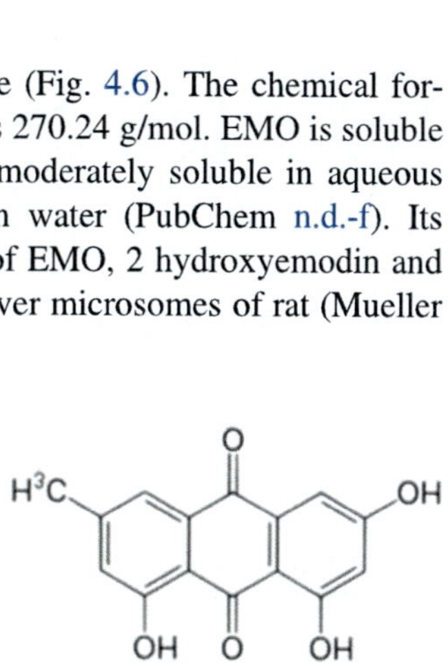

Fig. 4.6 Chemical structure of emodin. (By Yikrazuul—Own work, Public Domain, https://commons.wikimedia.org/w/index.php?curid=9291019)

4.6.2 Occurrence

EMO has been reported to occur in weevil-damaged chest nuts (Wells et al. 1975; Gross et al. 1984), leaves of aloe vera and sugar cane grass juice (De Saeger et al. 2016) and in feeds (Matthees 1983).

4.6.3 Toxicity

Emodin is a genotoxic, cytotoxic, immunosuppressive compound. It has been found to induce apoptosis in mouse blastocysts (Chang et al. 2012). The symptoms of emodin toxicity in cockerels included loss of appetite, acute epidermal irritation around the cloaca, genital debilitation, and death within 5 days of administration (Wells et al. 1975). EMO has been found to cause embryo toxicity in zebrafish (He et al. 2012). It can decrease the production of IL-1 and IL-2 and induce apoptosis in T cells and HK-2 cell lines. The long-term (2 years) exposure of EMO was observed to induce nephrotoxicity in Sprague-Dawley rats (Wang et al. 2008a, b; Liu et al. 2009; Gruber-Dorninger et al. 2017). Its hepatotoxicity has been demonstrated in mice. A very high dose (1–3 g/kg BW/day) of emodin has been found to develop melanosis in mice (Zheng et al. 2014). In 1981, Bachmann and Schlatter reported ADME properties of emodin in rats (Bachmann and Schlatter 1981). EMO has low bioavailability and mutagenicity (Wehner et al. 1979; Dong et al. 2016). However, a number of pharmacological benefits of emodin have been documented (Sougiannis et al. 2021).

4.7 Enniatins

Enniatins (ENNs) are produced by *Fusarium oxysporum*, *F. avenaceum*, *F. tricinctum*, *F. acuminatum*, and *F. sambucinum* (Jestoi 2008). ENNs are known as “minor mycotoxins”.

4.7.1 Chemical Structure

Ennatins are cyclic hexadepsipeptides having three D-hydroxyisovaleric acid and three *N*-methylated amino acids. Although 29 analogues (Table 4.1) of ennatin are identified, four of them, namely ENN A (Fig. 4.7), ENN A_1, ENN B, and ENN B_1 have been found frequently. The minor variants like ENN C, ENN D, ENN E, and ENN F have also been detected in *Fusarium*-contaminated cereals and cereal-based products (Altomare et al. 2021). The molecular weight of ENN B is 639.8 and its

Table 4.1 Analogs of enniatins (Hashizume and Nishimura 2008)

S. No.	Enniatins	R_1	R_2	R_3	R_4
1.	ENN A	$CH(CH_3)CH_2CH_3$	$CH(CH_3)CH_2CH_3$	$CH(CH_3)CH_2CH_3$	CH_3
2.	ENN B	$CH(CH_3)_2$	$CH(CH_3)_2$	$CH(CH_3)_2$	CH_3
3.	ENN B_1	$CH(CH_3)_2$	$CH(CH_3)_2$	$CH(CH_3)CH_2CH_3$	CH_3
4.	ENN B_2	$CH(CH_3)_2$	$CH(CH_3)_2$	$CH(CH_3)_2$	H
5.	ENN C	$CH_2CH(CH_3)_2$	$CH_2CH(CH_3)_2$	$CH_2CH(CH_3)_2$	CH_3
6.	ENN D	$CH(CH_3)_2$	$CH(CH_3)_2$	$CH_2CH(CH_3)_2$	CH_3
7.	ENN E	$CH(CH_3)_2$	$CH_2CH(CH_3)_2$	$CH(CH_3)CH_2CH_3$	CH_3
8.	ENN F	$CH_2CH(CH_3)_2$	$CH(CH_3)CH_2CH_3$	$CH(CH_3)CH_2CH_3$	CH_3
9.	ENN G	$CH_2CH(CH_3)_2$	$CH(CH_3)_2$	$CH_2CH(CH_3)_2$	CH_3
10.	ENN J_1	$CH(CH_3)_2$	$CH(CH_3)_2$	CH_3	CH_3
11.	ENN J_2	$CH(CH_3)_2$	$CH(CH_3)CH_2CH_3$	CH_3	CH_3
12.	ENN J_3	$CH(CH_3)_2$	CH_3	$CH(CH_3)CH_2CH_3$	CH_3
13.	ENN H	$R=CH(CH_3)_2$			
14.	ENN I	$R=CH(CH_3)CH_2CH_3$			

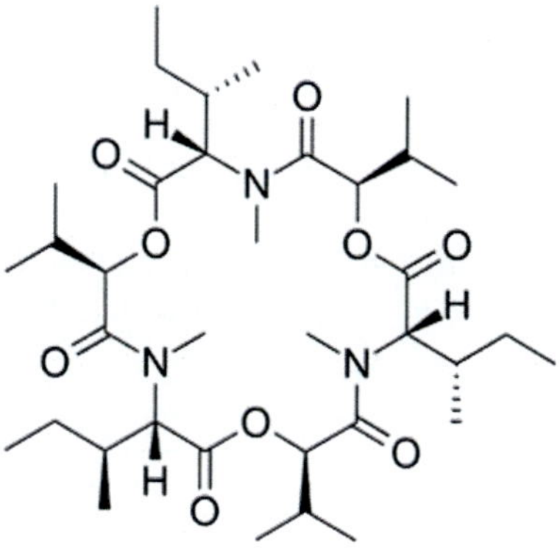

Fig. 4.7 Chemical structure of enniatin A. (Adapted from Ed (Edgar181)—Own work, Public Domain, https://commons.wikimedia.org/w/index.php?curid=15530620)

molecular formula is $C_{33}H_{57}N_3O_9$. The IUPAC name of ENN B is (3S,6R,9S,12R,15S,18R)-4,10,16-trimethyl-3,6,9,12,15,18-hexa(propan-2-yl)-1,7,13-trioxa-4,10,16-triazacyclooctadecane-2,5,8,11,14,17-hexone (PubChem n.d.-g).

4.7.2 Occurrence

ENNs have frequently been detected in processed grains including rye, oats, barley, wheat (Prosperini et al. 2017), and processed food products like pasta, wheat, corn, and barley couscous and semolina (Zinedine et al. 2017). ENN B has also been found to occur in beer samples and in dried distillers' grains with soluble (DDGS) (Habler et al. 2017; Granby et al. 2012).

4.7.3 Toxicity

ENNs have low toxicity in comparison to other minor mycotoxins produced by *Fusarium* spp. They are non-hazardous to humans and animals (EFSA 2014). However, ENN B may show synergistic/additive toxicological effects with other ennatins or other *Fusarium* mycotoxins (Fernández-Blanco et al. 2016). The mixture of ennatins has a cytotoxic effect on several cell lines (Prosperini et al. 2017). Ennatin B can induce apoptosis by cell cycle disruption (Gammelsrud et al. 2012) and has been shown to exert adrenal toxicity (Kalayou et al. 2015). The mitochondriotoxic properties of ennatins have also been documented (Tonshin et al. 2010). The in vivo studies in mice reported bioaccumulation of ENN B in lipophilic tissues without causing acute toxicity (Rodríguez-Carrasco et al. 2016). Moreover, the high bioavailability of ennatin B was reported in swine (Devreese et al. 2013) whilst its tissue accumulation was observed in broiler chickens (Fraeyman et al. 2016).

4.8 Fusaric Acid

Fusaric acid (FA) is an emerging mycotoxin and a secondary metabolite produced by *Fusarium moniliforme*, *F. heterosporum*, *F. oxysporum*, *F. crookwellense*, *F. verticilliodes*, *F. proliferatum*, *F. napiformes*, *F. subglutinans*, and *F. fujikuroi* (Bacon et al. 1996; Desjardins 2006).

4.8.1 Chemical Structure

FA is 5-butylpicolinic acid, a derivative of picolinic acid (Fig. 4.8). It is an off white to yellowish crystalline compound and a chelating agent. The 5-butyl side chain increases the lipophilicity of FA, thereby facilitating its entry into the cell (Arumugam et al. 2021). Biosynthesis of fusaric acid involves two enzymes, pyridoxal 5′-phosphate (PLP)-dependent enzyme (Fub 7) and flavin mononucleotide-dependent oxidase (Fub 9) (Hai et al. 2020). *Burkholderia ambiferia* is capable of detoxifying FA (Simonetti et al. 2018; PubChem n.d.-h).

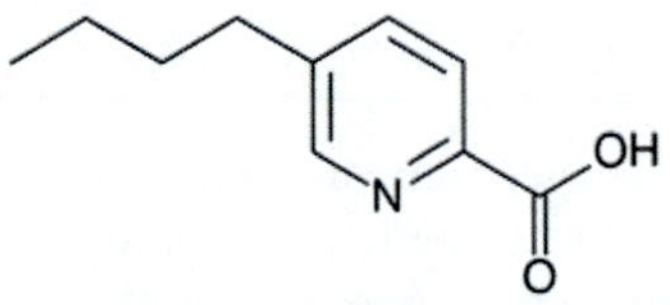

Fig. 4.8 Chemical structure of fusaric acid. (Own work assumed (based on copyright claims)., Public Domain, https://commons.wikimedia.org/w/index.php?curid=695342)

4.8.2 Occurrence

Corn and corn-based food products and feed, swine feed, poultry, and livestock feeds (Smith and Sousadias 1993; Porter et al. 1995; Bacon et al. 1996).

4.8.3 Toxicity

FA has phytotoxicity and was found to accelerate wilt in banana and tomato plants (Selim and El-Gammal 2015). It can suppress cell-mediated immunity in chickens (Chu et al. 1993). Fusaric acid has teratogenic effects on zebra fish (Yin et al. 2015). The symptomatology of FA-toxicity in broilers includes elevation in the levels of epinephrine and norepinephrine and reduction in the weight of kidneys (Nagatsu et al. 1970). The acute dosing of FA caused neuro-chemical changes and vomiting while the combined effect of FA and trichothecenes led to feed refusal in pigs (Smith and MacDonald 1991). The exposure of fusaric acid lowered the levels of epinephrine and norepinephrine in rabbits (Nagatsu et al. 1970) but raised serotonin levels in rats (Chaouloff et al. 1986). It can alter the neurotransmitters of the brain and pineal gland in rats (Porter et al. 1995). In 2013, Devaraja et al. observed biphasic dose-dependent effect of FA on the coagulation of human plasma. They also reported its ability to activate platelet aggregation in humans (Devaraja et al. 2013). FA is considered as a precursor of beta-blocker and can have hypotensive effects on cats, dogs, rabbits, rats, and humans (Hidaka et al. 1969; Velasco et al. 1975). IC_{50} and the toxic effects of fusaric acid on different cell lines have been shown in Table 4.2.

The contamination of fusaric acid poses potential health risks to humans and animals. Its toxicity and co-toxicity (with DON) should be appropriately managed.

Table 4.2 Toxic effects of fusaric acid on different cell lines (Vesonder et al. 1993; Abdul et al. 2016)

S. No.	Mammalian cell lines	IC_{50}	Toxic effects
1.	Kidney (dog)	10 µg/ml	Cytotoxic
2.	Fibroblast cells of Chinese hamster	10 µg/ml	Cytotoxic
3.	Rat hepatoma cells	50 µg/ml	Cytotoxic
4.	WI-38 fibroblast (human)	500 µm	Cell proliferation suppression
5.	Colorectal adenocarcinoma (human)	500 µm	Cell proliferation suppression
6.	HePG2 (human) hepatocellular carcinoma cell line	–	Mitochondrial stress, apoptosis, necrosis, oxidative stress

4.9 Fusaproliferin

Fusaproliferin (FUS) was previously known as proliferin but in 1995, this *Fusarium* metabolite was renamed as fusaproliferin (Randazzo et al. 1993; Ritieni et al. 1995). It was designated as the "emerging mycotoxin" in the year 2008 (Jestoi 2008). Although FUS mycotoxin was discovered as a toxic metabolite of *Fusarium proliferatum*, but can also be produced by *F. subglutinans*, *F. fujikuroi*, *F. pseudocircinatum*, *F. udum*, *F. begonice*, *F. concentricum*, *F. solani*, *F. antophilum*, *F. bulbicola*, *F. succisae*, and *Gibberella* spp. (Santini et al. 1996; Fotso et al. 2002; Moretti et al. 2007; Hoque et al. 2018). FUS occurs (in 1:3 ratio) with its deacetylated analogue, deacetyl-FUS (Moretti et al. 2007).

4.9.1 Chemical Structure

Fusaproliferin is a bicyclic sesterterpene. Its chemical structure consists of a new ring skeleton with 4 chiral atoms and 4 C=C (Santini et al. 1996). The molecular weight of FUS is 444.6 and the molecular formula is $C_{27}H_{40}O_5$ (Fig. 4.9). The IUPAC name of FUS is [(2S)-2-[(1S,3E,7E,11S,12E,15R)-11,17-dihydroxy-1,4,8,12-tetramethyl-18-oxo-16 bicyclo[13.3.0]octadeca-3,7,12,16-tetraenyl]propyl] acetate (PubChem release 2019). It is produced by terpene intermediates through an isoprenoid pathway (Santini et al. 1996).

4.9.2 Occurrence

The natural occurrence of fusaproliferin has been detected in cereal samples from Valencian community, Spain (Meca et al. 2010); in domestic rice from Gilan and Mazandaran Provinces of Iran; in organic cereals and other products from local markets of Italy (Juan et al. 2013); in wheat, barley, and maize from Rabat Sale area of Morocco (Zinedine et al. 2011); in maize samples from Serbia (Jajić et al. 2019); in contaminated maize from Iowa, USA (Munkvold et al. 1998); in maize samples from Transkei region, South Africa (Shephard et al. 1999); in rice samples collected

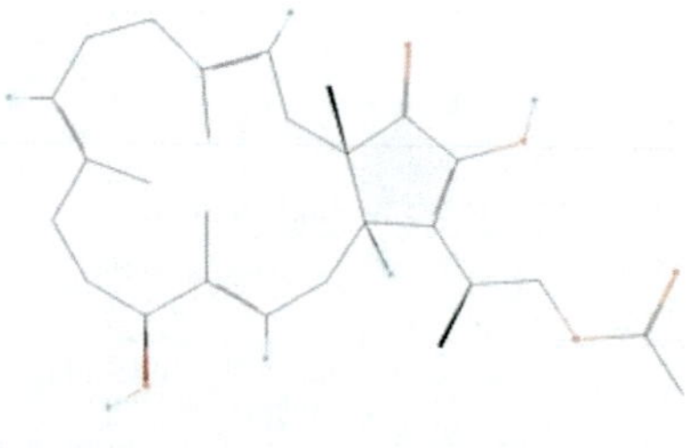

Fig. 4.9 Chemical structure of fusaproliferin. (Pubchem 2019)

from Morocco (Sifou et al. 2011); in wheat, oats, rye, and barley from Finland (Jestoi et al. 2004); in an Italian grain product "farro" (Castoria et al. 2005) and in infant cereals from Morocco (Mahnine et al. 2011).

4.9.3 *Toxicity*

FUS has been found toxic to larvae of brine shrimp *Artemia salina* and moderately toxic to *Spodoptera frugiperda* (SF-9) cell line (Logrieco et al. 1996). It has been shown to cause teratogenic changes such as macrocephaly, cephalic dichotomy, and limb asymmetry in the chick embryo (Ritieni et al. 1997). The LC_{50} value of fusaproliferin to human B-lymphocyte cell line (IARC/LCL170) was calculated to be 55 μM (Logrieco et al. 1996).

Despite its wide occurrence in agricultural commodities, various aspects of fusaproliferin toxicity like symtomatology, permissible limits, bio-monitoring and management are still not determined. Therefore, more research should be done in this direction.

4.10 Gliotoxin

Gliotoxin (GT) is a naturally occurring sulphur-containing mycotoxin. It is mainly produced by *Aspergillus fumigatus*. However, *Eurotium chevalieri*, *Gliocladium fimbriatum*, and several species of *Penicillium* and *Trichoderma* are also known to produce GA (El-Bondkly 2014). It was first isolated from *Gliocladium fimbriatum* in 1930, thus named as gliotoxin (Johnson et al. 1943). GA has been isolated from the sera of patients suffering from invasive aspergillosis and is considered as one of the virulence factors of *A. fumigatus* (Kwon-Chung and Sugui 2009). The inactive derivative of gliotoxin, bmGT (bis methyl gliotoxin) has also been found in *A. fumigatus* (Matxalen et al. 2018). *A. fischeri*, a non-pathogenic close relative of *A. fumigatus*, has also been shown to biosynthesize gliotoxin (Knowles et al. 2020).

4.10.1 *Chemical Structure*

Gliotoxin (GT) is a naturally occurring epipolythiodioxopiperazine with the chemical formula $C_{13}H_{14}N_2O_4S_2$ (Fig. 4.10). GT occurs in oxidized and reduced structural forms (Nirwan and Kakkar 2019). Its molecular weight is 326.4. Its IUPAC name is (1R,7S,8S,11R)-7-hydroxy-11-(hydroxymethyl)-15-methyl-12,13-dithia-9,15-diazatetracyclo[9.2.2.01,9.03,8]pentadeca-3,5-diene-10,14-dione (PubChem n.d.-i).

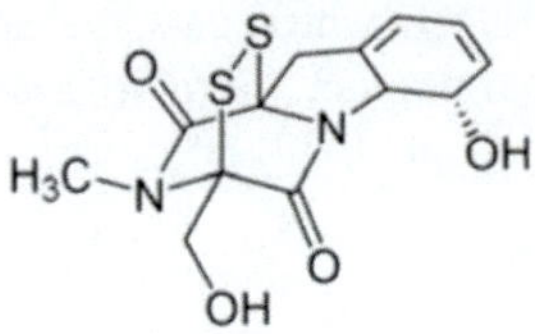

Fig. 4.10 Chemical structure of gliotoxin. (By Yikrazuul—Own work, Public Domain, https://commons.wikimedia.org/w/index.php?curid=1440747)

4.10.2 Occurrence

The main source of gliotoxin, *A. fumigatus*, is a general contaminant. Thus, GA has a ubiquitous occurrence in the environment.

4.10.3 Toxicity

Gliotoxin has immunosuppressive effects. It can inhibit CTLs-mediated FasL-dependent and perforin killing pathways (Kataoka and Nagai 2002). GT exerts cytotoxicity (through apoptosis), ROS production, anti-angiogenesis, and genotoxicity (Scharf et al. 2016).

Despite its contribution to the pathogenicity of *Aspergillus fumigatus*, gliotoxin is useful in crop protection and in the treatment of hepatic necrosis and cystic fibrosis. Thus, biosynthetic methods should be used to alleviate its toxic potential.

4.11 Moniliformin

Moniliformin (MON) was isolated from extracts of *Fusarium moniliforme* in 1972 (Cole et al. 1973). Several species of *Fusarium* like *F. subglutinans*, *F. proliferatum*, *F. tricinctum*, *F. avenaceum*, *F. fujikuroi*, *F. pesudonygamai*, *F. nygamai*, and *F. temperatum* can synthesize moniliformin (Leslie et al. 2005; Desjardins et al. 2006; Scauflaire et al. 2012). Apart from these, MON can also be produced by *Penicillium melanoconidium* (Hallas-Moeller et al. 2016). However, there is no record of a natural occurrence of MON-derivatives (EFSA 2018).

4.11.1 Chemical Structure

Chemically, MON is a sodium or potassium salt of 1-hydroxycyclobut-1-ene-3,4-dione (Springer et al. 1974; Uhlig et al. 2004). The molecular weight of

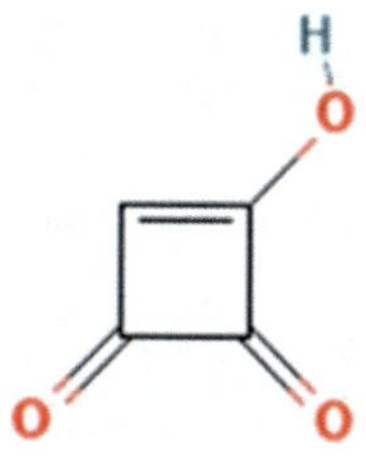

Fig. 4.11 Chemical structure of moniliformin. (PubChem n.d.-j)

moniliformin is 98.6 and its molecular formula is $C_4H_2O_3$ (PubChem n.d.-j) (Fig. 4.11). MON is highly soluble in water and is also known as semisquaric acid (Diaz Toro et al. 2015). As a thiamine inhibitor, moniliformin has been shown to inhibit the activity of thiamine-dependent enzymes like pyruvate dehydrogenase and pyruvate decarboxylase (Pirrung et al. 1996).

4.11.2 Occurrence

Natural contamination of MON has been reported in cereal grains (oats, barley, wheat) and poultry feed from Finland, Slovakia, and Lesotto. Its contamination in maize, wheat, and ray has been recorded from European Union, China, the United Kingdom, Switzerland, The Netherlands, Poland, and Italy (Jestoi 2008; van Asselt et al. 2012; Scarpino et al. 2013).

4.11.3 Toxicity

The phytotoxicity of MON was reported in corn, wheat, and tobacco (Cole et al. 1973). In channel fish, a low dose (0.8 mg/kg/day) of MON caused reduction in weight gain (EFSA 2018). Its presence in feed reduced the weight gain and feed refusal in catfish, poultry, and in swine (Fremy et al. 2019). MON has been seen to cause heart failure, respiratory distress, and muscular weakness in farm animals. The performance and immunity of the animals have also been affected (Fremy et al. 2019). The moniliformin-induced cardiac muscle disruption has been observed in Japanese quails and broilers (Javed et al. 2005; Sharma et al. 2012). MON is highly toxic to chickens. The mild form of diffuse mineralization was observed in the renal tubules of birds fed with MON (Fremy et al. 2019). Its intraperitoneal and oral administration caused neonatal mortality in mink (Jestoi 2008). MON toxicity has not been considered hazardous in farmed fish, farmed rabbits, cats, dogs, horses, and ruminants (EFSA 2017). In a laboratory setting, MON has been found less toxic to mice than in chicks (Burmeister et al. 1979). The exposure of MON induced chromosomal aberration in the primary hepatocytes of rats and genotoxicity in the leucocytes of humans. As toxic effects of moniliformin in humans have not been reported yet, its human acute or chronic health-based guidance value (HBGV) has not been determined till date (EFSA 2017).

Since MON is extremely toxic to poultry birds, the monitoring of poultry feed should be done regularly.

4.12 Mycophenolic Acid

Mycophenolic acid is a well-known antirheumatic drug and is widely used for immunosuppression during organ transplantation (Cholewinski et al. 2010). It was the first time isolated from *Penicillium brevicompactum* in 1893 (Bentley 2000). MPA can also be produced by *P. carneum*, *P. echinulatum*, *P. roqueforti*, *P. stoloniferum*, and *Byssochlamys nivea* (Anderson et al. 1988; Frisvad et al. 2006).

4.12.1 Chemical Structure

Mycophenolic acid (MPA), a member of the class of 2-benzofurans, is (4E)-6-(4-hydroxy-6-methoxy-7-methyl-3-oxo-5-phthalanyl)-4-methyl-4-hexenoic acid (Fig. 4.12). Its molar mass is 320.34 g/mol. The chemical formula of MPA is $C_{17}H_{20}O_6$ (PubChem n.d.-k).

4.12.2 Occurrence

MPA was detected in ginger, Roquefort cheese and German blue-white soft cheese samples, red wine and silage (Usleber et al. 2008; Sørensen et al. 2008). The presence of MPA was also observed in dried ham and liver pâté contaminated with *Penicillium brevicompactum* (Sørensen et al. 2010).

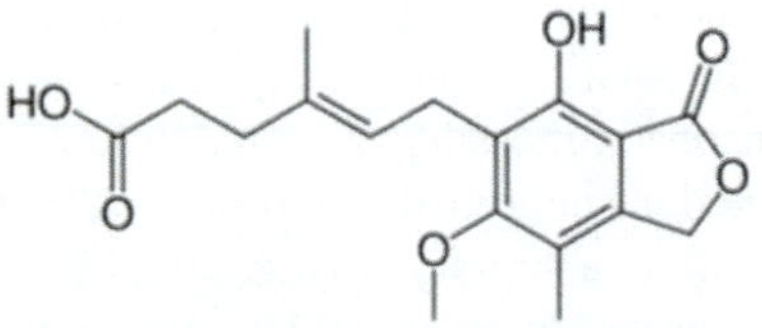

Fig. 4.12 Chemical structure of mycophenolic acid. (By The original uploader was Meodipt at English Wikipedia. Public Domain, https://commons.wikimedia.org/w/index.php?curid=8401309)

4.12.3 Toxicity

MPA has low toxicity for rats (LD_{50} approximately 450 mg/kg bw) and mice (LD_{50} approx. 1900 mg/kg bw) (Sweeney et al. 1972; Gruber-Dorninger et al. 2017). However, its high doses caused anaemia and growth retardation in rats. The low doses of MPA caused diarrhoea, enteritis, and anorexia in dogs, while exposure to high doses led to fatalities (Sweeney et al. 1972). The common side effects of its sodium salt and ethyl ester (mycophenolate mofetil) in humans include urinary anaemia, GI tract intolerance, insomnia, dyspepsia, and leukopenia. During pregnancy, its use may cause miscarriage and congenital malformations (Drugs.com 2021).

4.13 Penicillic Acid

Penicillic acid (PA), a small lactone, was first isolated from *Penicillium puberulum* Bainier (Alsberg and Black 1913). Later on, Birkinshaw et al. (1936) isolated PA from *P. cyclopium*. Presently, 24 *Penicillium* species from nine sections viz. *Turbatumand*, *Brevicompacta*, *Roquefortorum*, *Canescentia*, *Lanata-Divaricata*, *Exilicaulis*, *Chrysogena*, *Fasciculata*, and *Ramosa* like *Penicillium brasilianum* (grass-associated mould), *Penicillium baarnense*, *Penicillium jamesonlandense* and *Penicillium egyptiacum* (soil-borne moulds) and *Penicillium bovifimosum* (dung-associated mould), 23 species of *Aspergillus* section *Circumdati*, and some strains of *Aspergillus ochraceus* have been found to produce penicillic acid (Frisvad 2018).

4.13.1 Chemical Structure

PA is (2Z)-3-methoxy-5-methyl-4-oxohexa-2,5-dienoic acid (Fig. 4.13). The chemical formula and molecular weight of penicillic acid are $C_8H_{10}O_4$ and 170.16 respectively (PubChem n.d.-l).

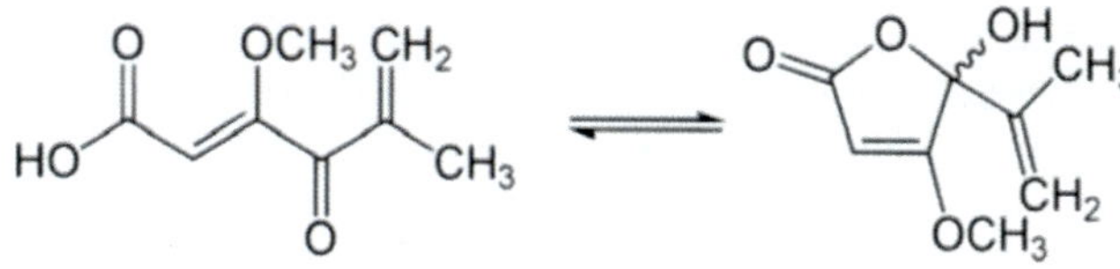

Fig. 4.13 Chemical structure of penicillic acid. (Adapted from Yikrazuul—Own work, Public Domain, https://commons.wikimedia.org/w/index.php?curid=7213094)

4.13.2 Occurrence

The high moisture content and low-temperature storage favour the production of penicillic acid in large quantities (Bianchini and Bullerman 2014). The natural occurrence of PA has also been detected in cassava and cereals (Bokhari and Flannigan 1996; Wareing et al. 2001). Moreover, the co-contamination of PA and terrestric acid was reported in apples, cheese, nuts, and rice as well (Yu et al. 2014).

4.13.3 Toxicity

PA, along with patulin, inhibits DNA, RNA, and protein synthesis. Penicillic acid is also capable of blocking the enzymes containing the SH group (Wilson 1976). The LD_{50} values of PA in mice for oral, intraperitoneal, and intravenous routes were found to be 600 mg/kg, 90 mg/kg, and 250 mg/kg respectively (Chan et al. 1984). It induces hepatotoxicity, cytotoxicity, and carcinogenicity in mice. PA is non-carcinogenic to humans thereby included in category 3 of the International Agency for Research on Cancer (IARC 2014). Nonetheless, it has been seen to cause chromosomal breakage and aberrations, thereby causing mutagenicity (Umeda et al. 1977). Penicillic acid and its derivative dihydropenicillic acid can slow down the activation of NF-kappa B (Tachibana et al. 2008). PA targets self-processing of caspase 8 (cap 8) by binding directly with its large subunit and inhibits the apoptosis induced by Fas ligand (Bando et al. 2003). The toxicological profile of PA is summarized in Table 4.3.

Table 4.3 Toxicological profile of penicillic acid

S. No.	Toxicity caused by PA	Animals/cell lines	Reference/s
1.	Phytotoxicity	–	Ismaiel and Papenbrock (2015)
2.	Mycotoxicosis	Broilers	Ghadevaru (2013)
3.	Cardiotoxic	Frog	Pandiyan et al. (1990)
4.	Toxic	Rat alveolar macrophage	Sorenson and Simpson (1986)
5.	Acute toxic	Rabbits	Cole and Cox (1981)
6.	Synergistic toxicity with ochratoxin A	(a) Bovine macrophage cell line	Oh et al. (2015)
		(b) Fall armyworm and corn ear worm	Dowd (1989)
		(c) Chickens	Micco et al. (1991)
7.	Balkan endemic nephropathy		Stoev (2017)

4.14 Phomopsins

The teleomorph of Phomopsis mould, Diaporthe toxica produces phomopsins in the contaminated lupin crop. The other producers of phomopsins are *D. woodii* and *P. emecis* (van Warmelo and Marasas 1972).

4.14.1 *Chemical Structure*

Phomopsins are hexapeptides. The IUPAC name of phomopsin A is (*E*)-2-[[(*E*)-2-[[(2*S*)-1-[(3*R*,4*S*,7*S*,10*S*,11*S*)-14-chloro-3-ethyl-11,15-dihydroxy-3-methyl-10-(methylamino)-6,9-dioxo-7-prop-1-en-2-yl-2-oxa-5,8-diazabicyclo[10.3.1]hexadeca-1(15),12(16),13-triene-4-carbonyl]-2,5-dihydropyrrole-2-carbonyl]amino]-3-methylpent-2-enoyl]amino]but-2-enedioic acid (Fig. 4.14). Its chemical formula is $C_{36}H_{45}ClN_6O_{12}$ whereas molecular weight is 789.2 (PubChem n.d.-m).

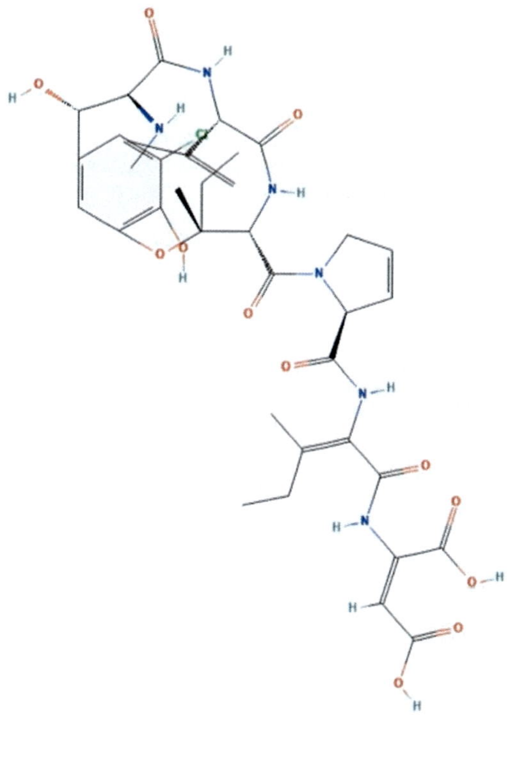

Fig. 4.14 Chemical structure of Phomopsin A

4.14.2 Toxicity

Out of five metabolites (namely A, B, C, D, and E), phomopsin A has the highest toxicity. Dietary exposure of phomopsin A-contaminated lupin fodder causes lupinosis in sheep and cattle. It is a potent hepatotoxin, thus affecting the liver the most. However, the toxic effects of phomopsin A have also been seen in the pancreas, hepatic lymph nodes, rumen, kidneys, and adrenal glands (www.alltech.com, Alltech n.d.). It is anti-mitotic and can inhibit microtubule assembly (Ludueña et al. 1989).

4.15 Sterigmatocystin

Sterigmatocystin (STC), a carcinogen, is produced by *Aspergillus versicolor*, *A. parasiticus*, *A. flavus*, *A. nidulans*, *A. ruber*, *A. sydowi*, *A. quadrilineatus*, *A. aureolatus*, *A. chevalieri*, *A. Amstelodami*, *A. creber*, *A. amoenus*, *A. fructus*, *A. griseoaurantiacus*, *A. subversicolor*, *A. pepii*, *A. protuberus*, *A. hongkongensis*, *A. cvjetkovicii*, *A. tennesseensis*, *A. jensenii*, *A. venenatus*, *A. puulaauensi* (Jakšić et al. 2017), and several other moulds like *Bipolaris*, *Penicillium*, *Emiricella*, and *Chaetomium* (Versilovskis and De Saeger 2010). The poor hygienic conditions, high humidity, and contaminated air of warehouses are responsible for the production of STC in stored grains (Dobson 2017). The fluctuation in temperature, low availability of water and micronutrients, damage caused by insect pests and other storage-related problems also influence STC production (Viegas et al. 2020).

4.15.1 Chemical Structure

STC is a heteropentacyclic organic compound having a xanthenes ring in its skeleton (PubChem n.d.-n). The molecular formula of sterigmatocystin is $C_{18}H_{12}O_6$ while the molecular weight is 324.3. The IUPAC name of STC is (3S,7R)-15-hydroxy-11-methoxy-6,8,20-trioxapentacyclo[10.8.0.02,9.03,7.014,19]icosa-1,4,9,11,14,16,18-heptaen-13-one (Fig. 4.15). It is insoluble in water and petroleum ether, hardly

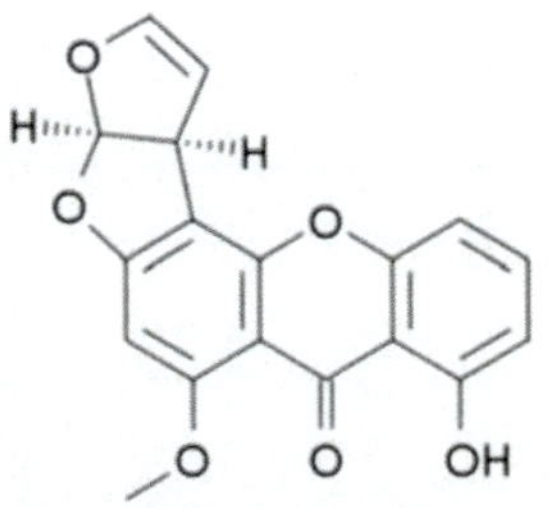

Fig. 4.15 Chemical structure of sterigmatocystin. (By Meodipt at en. wikipedia. Public Domain, https://commons.wikimedia.org/w/index.php?curid=16732461)

soluble in diethyl ether, methanol, and ethyl alcohol, while soluble in organic solvents like benzene, acetone, and ethyl acetate (Tabata 2002). Sterigmatocystin is a precursor of aflatoxins and is very similar to AFB_1.

4.15.2 Occurrence

Sterigmatocystin seldom occurs in nature. However, its occurrence in stored grains and cheese was reported (Tabata 2011). A low concentration of STC has been detected in the rind of hard Dutch cheese, mouldy wheat, and green coffee (Groopman et al. 2013). The contamination of sterigmatocystin has also been found in wheat flour, buckwheat flour, rye flour, corn flour, barley products, white sorghum, Job's tear products, and azuki bean (Yoshinari et al. 2019; Tabata 2022), Gouda cheese, Edam cheese and hard cheese (Dobson 2017), almonds (Yassin et al. 2013), peanuts, cattle feed (Versilovskis and De Saeger 2010), soybean, silage, sunflower cake, piglet feed mixture (Warth et al. 2012; El-Shanawany et al. 2005; Kovalenko et al. 2011), spices, bread and beer (Nieto et al. 2018). Apart from food and feed, STC was also isolated from wall paper and carpet dust of the damp indoor environment (Gao et al. 2015).

4.15.3 Toxicity

The exposure of STC reduced the crude protein and body growth in cat fish and carp (Abdelhamid 1988) whereas it induced chromosomal damage in Nile tilapia (*Oreachromis niloticus*) (Abdel-Wahhab et al. 2005). STC has potential mutagenicity and can cause hepatotoxicity in male rats (Tabata 2011), mice, guinea pigs, and monkeys (Viegas et al. 2020). Sterigmatocystin-induced renal tubule necrosis, haemorrhage, and hyaline degeneration were also observed in rats and monkeys (EFSA 2019). The indoor exposure of this mycotoxin induced severe non-specific inflammatory responses in rats (Miller et al. 2010; Piecková 2015). STC contamination in feeding stuff caused bloody diarrhoea and death in dairy cattle (Versilovskis and De Saeger 2010). Sterigmatocystin is classified as a 2B carcinogen by the IARC due to its possible carcinogenicity to humans. The high incidences of gastric and liver cancer in several areas of China have been correlated with STC toxicity (Chen and Inbaraj 2016). A number of in vitro studies have been conducted to observe the effects of STC on different cell lines as well (Table 4.4).

Sterigmatocystin is a potent genotoxic and carcinogen. In spite of this, only a few countries such as Slovakia, the Czech Republic, the EU, and the USA have safety regulations for this. Since STC contaminates grains during storage, therefore, better post-harvest management of crops is highly recommended. Moreover, sensitive methods should be developed for the detection of STC in food and feed.

Table 4.4 Toxicological effects of STC on different cell lines

S. No.	Cell line/cell assay	Toxic effects	Reference/s
1.	Bacterial cell assay	Mutagenic	Mori et al. (1986)
2.	Mammalian cell assay	Mutagenic	Curry et al. (1984)
3.	HePG2 cells	Genotoxicity	Gao et al. (2015)
4.	Human oesophageal epithelial cells (Het-1A)	DNA damage G2 phase arrest (possible risk for oesophageal cancer)	Wang et al. (2015)
5.	Human pulmonary cells	DNA damage and genomic instability	Huang et al. (2014)
6.	Ovarian hamster cell (CHO-K1 cell)	Cytotoxicity	Zouaoui et al. (2016)

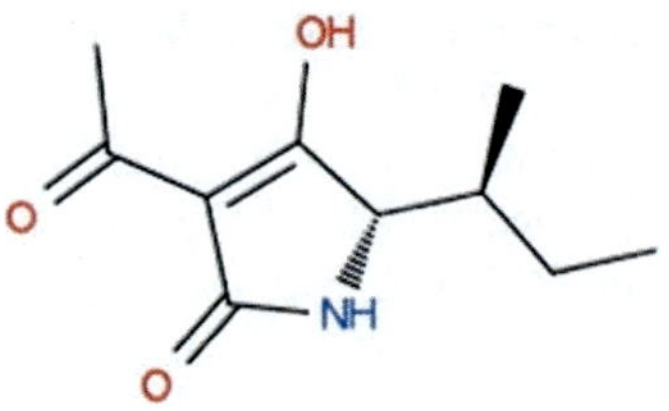

Fig. 4.16 Structure of tenuazonic acid. (By Mfernflower—Own work, Public Domain, https://commons.wikimedia.org/w/index.php?curid=43101444)

4.16 Tenuazonic Acid

Tenuazonic acid is considered to have the highest toxicity amongst the *Alternaria* mycotoxins (Weidenbörner 2001). TeA is a toxic metabolite produced by *Alternaria* spp., *Phoma sorghina*, and *Pyricularia oryzae* (Iwasaki et al. 1972; Steyn and Rabie 1976). Tenuazonic acid is predominantly produced by *Alternaria* spp., for instance, *A. arborescens*, *A. alternata*, *A. capsicianui*, *A. citri*, *A. japonica*, *A. kikuchiana*, *A. mali*, *A. solani*, *A. oryzae*, *A. porri*, *A. radicina*, and *A. tenuissima*. The other TeA producer fungi are *Phoma sorghina* and *Pyricularia oryzae* (Umetsu et al. 1974, 1972; Steyn and Rabie 1976; Ukwuru et al. 2017). The optimum temperature and minimum a_w for the production TeA are reported to be 25 °C and 0.90 a_w (Ostry 2008).

4.16.1 Chemical Structure

Tenuazonic acid is colourless, viscous oil and is a monobasic acid with pK_a 3.5. The chemical formula of TeA is 3-acetyl-5-sec-butyl-4-hydroxy-3-pyrrolin-2-one; its molecular weight (MW) is 197 and the molecular formula is $C_{10}H_{15}O_3N$ (Ostry 2008; The Merck Index 2018) (Fig. 4.16). TeA is soluble in organic solvents including petroleum ether but is sparingly soluble in water (https://www.rsc.org/

Merck-Index/monograph/m10562/tenuazonic%20acid?q=authorize). It is ketonic in nature. Its rotation in methanol is −124°. The rotation of TeA in crude form slowly becomes less negative and crystallizes when kept standing for a long time. The purified form of TeA is dextrorotatory. It titrates sharply as a monobasic acid, contains no methoxyl group, and forms a semicarbazone. When boiled with an aqueous alkali, it is converted to an iso acid form (Shigeura 1967). It forms complexes with calcium, magnesium, copper, iron, and nickel ions. Tenuazonic acid is usually stored as its copper salt (Kumari and Tirkey 2019).

4.16.2 Occurrence

Tenuazonic acid is predominantly present in tomatoes and its pulp, puree, sauce, and ketchup. It is known to be present in a number of cereals and their products which include maize, millet, wheat, sorghum, rice, beer, bran (wheat bran), cassava, flakes (corn flakes, oat flakes), flour (rye flour, wheat flour), and grit (maize grits). It is also found in fruits such as apple, mandarin, melon, olive, dried fruit, and spices such as coriander, cumin, curcuma, curry, ginger, paprika, pepper, and in the juice of apple, apricot, banana, blackcurrant, cherry, grape, orange, pear, pomegranate, strawberry, and tomato. The occurrence of TeA has also been detected in sesame and sunflower seeds (Ostry 2008; Kumari and Tirkey 2019).

4.16.3 Toxicity

TeA acts as an inhibitor of photosystem II and obstructs the transfer of electrons from Q_A to Q_B in photosystem II (Chen et al. 2007). *A. alternata* is known to cause infection in host plants with the help of TeA as it has been reported as a chief virulence factor in causing infection. TeA toxin recognizes the host, induces appressoria for infection in the host plant, and maintains the level of ROS until the infection occurs completely (Meronuck et al. 1972).

The toxicity of TeA has been studied in several animals like mice, chicken, and dogs (Steyn and Rabie 1976; Griffin and Chu 1983). It causes haemorrhages in several organs in dogs when given at a dose of 10 mg/kg body weight. Sub-acute toxicity was observed in chickens when fed at a dose of 10 μg/g of feed. In chickens, the sub-lethal to lethal doses of tenuazonic acid caused an increase in internal haemorrhage, suppression in weight gain, and reduction in feed efficiency (Griffin and Chu 1983; Woody and Chu 1992). The oral administration of TeA (25 mg/kg bw/day) for 10 months in mice induced precancerous changes in the oesophageal mucosa (Steyn and Rabie 1976). TeA inhibits protein biosynthesis and is biologically active exerting antitumor and antibiotic activities (Steyn and Rabie 1976; Shephard et al. 1991; Weidenbörner 2001; Schobert and Schlenk 2008). It was found to be associated with mycotoxicosis, Onyalai, a haematological disorder observed in the African population (Lurie et al. 1969; Steyn and Rabie 1976). The clinical feature of the disease is haemorrhagic bullae on the mucosa of the oronasopharynx. The haemorrhage from ruptured bullae, epistaxis, or gastrointestinal bleeding may become severe and fatal (Ostry 2008).

Shigeura and Gordon (1963) investigated the biological activity of tenuazonic acid, in vivo with intact rats and in vitro with Ehrlich ascites tumour and rat liver cells and demonstrated that TeA inhibited the incorporation of amino acids into proteins by interacting with the peptidyltransferase in ribosomes (Chu 2003). In cell-free systems, TeA-induced inhibition of protein synthesis occurs by the suppression of newly formed microsomal proteins (Shigeura and Gordon 1963). In another study carried out by Kumari and Singh (2021), TeA was reported to cause multiple organ toxicity triggered by oxidative stress (Kumari and Singh 2021).

Regulations regarding the limit of *Alternaria* toxins have not yet been set. However, the European Union is most likely to set up allowable limits for *Alternaria* toxins (EFSA 2011; Gil-Serna et al. 2014a; Arcella et al. 2016).

4.17 Conclusion

Emerging mycotoxins are hidden threats. They are minor toxins produced predominantly by mycotoxigenic filamentous moulds. The microfungus *Stachybotrys chartarum* and the yeast-like fungus *Aureobasidium pullulans* are indoor contaminants and can produce toxic metabolites. The verrucarin J, trichoverrols A, B, satratoxins F, G, H, and roridin E are produced by *S. chartarum*. The stachybotryotoxins are the causative agents of sick building syndrome and eye infection in humans. The feeding or bedding of contaminated straw causes contact dermatitis in broilers and swine. The stachybotry-intoxication induces erythema, oedema of lips, tongue, and nose, and gastroenteritis in the horses (Dalefield 2017). Likewise, the toxicity of *A. pullulans* metabolite is also evident in experimental mice (Kumari and Singh, unpublished data). Although acute toxicity of emerging mycotoxins is not reported in humans, nonetheless the possibility of carry-over toxicity cannot be denied. Therefore, more attention should be given to the detection, structural elucidation, mode of action, and toxicological profile of these mycotoxins.

References

Abdelhamid AM (1988) Effects of sterigmatocystin contaminated diets on fish performance. Archiv für Tierernaehrung 38:833–846

Abdel-Wahhab MA, Hasan AM, Aly SE, Mahrous KF (2005) Adsorption of sterigmatocystin by montmorillonite and inhibition of its genotoxicity in the Nile tilapia fish (*Oreachromis nilaticus*). Mutat Res 582(1–2):20–27

Abdul NS, Nagiah S, Chuturgoon AA (2016) Fusaric acid induces mitochondrial stress in human hepatocellular carcinoma (HepG2) cells. Toxicon 119:336–344

Abia WA, Warth B, Sulyok M, Krska R, Tchana AN, Njobeh PB, Dutton MF, Moundipa PF (2013) Determination of multimycotoxin occurrence in cereals, nuts and their products in Cameroon by liquid chromatography tandem mass spectrometry (LC-MS/MS). Food Control 31:438–453

Agosti G, Birkinshaw JH, Chaplen P (1962) Studies in the biochemistry of micro-organisms. 112 Anthraquinone pigments of strains of *Cladosporium fulvum* Cooke. Biochem J 85:528–530

Alltech (n.d.). https://www.alltech.com/en-gb/blog/emerging-mycotoxins-what-do-we-know-so-far

Alsberg CI, Black OF (1913) Contribution to the study of maize deterioration; biochemical and toxicological investigations of *Penicillium puberulum* and *Penicillium stoloniferum*. USDA Bureau Plant Indus Bull 270:7–48

Altomare C, Logrieco AF, Gallo A (2021) Mycotoxins and mycotoxigenic fungi: risk and management. a challenge for future global food safety and security. In: Zaragoza O, Casadevall A (eds) Encyclopedia of mycology. Elsevier, Amsterdam, pp 64–93. https://doi.org/10.1016/B978-0-12-819990-9.00032-9

Anderson HA, Bracewell JM, Fraser AR, Jones D, Robertson GW, Russell JD (1988) 5-Hydroxymaltol and mycophenolic acid, secondary metabolites from *Penicillium echinulatum*. Trans Br Mycol Soc 91(4):649–651. https://doi.org/10.1016/S0007-1536(88)80040-8

Arcella D, Eskola M, Ruiz JAG (2016) Dietary exposure assessment to *Alternaria* toxins in the European population. EFSA J 14:4654

Arumugam T, Ghazi T, Abdul NS, Chuturgoon AA (2021) A review on the oxidative effects of the fusariotoxins: Fumonisin B1 and fusaric acid. In: Patel VB, Preedy VR (eds) Toxicology. Academic Press, London, pp 181–190. https://doi.org/10.1016/B978-0-12-819092-0.00019-4. ISBN 9780128190920

Ashley JN, Hobbs BC, Raistrick H (1937) Studies in the biochemistry of micro-organisms: the crystalline colouring matters of *Fusarium culmorum* (W. G. Smith) Sacc. and related forms. Biochem J 31(3):385–397. https://doi.org/10.1042/bj0310385

Bachmann M, Schlatter C (1981) Metabolism of [14C]emodin in the rat. Xenobiotica 11(3): 217–225. https://doi.org/10.3109/00498258109045294

Bacon CW, Porter JK, Norred WP, Leslie JF (1996) Production of fusaric acid by *Fusarium* species. Appl Environ Microbiol 62(11):4039–4043. https://doi.org/10.1128/aem.62.11.4039-4043.1996

Bando M, Hasegawa M, Tsuboi Y, Miyake Y, Shiina M, Ito M, Handa H, Nagai K, Kataoka T (2003) The mycotoxin penicillic acid inhibits Fas ligand-induced apoptosis by blocking self-processing of caspase-8 in death-inducing signaling complex. J Biol Chem 278(8):5786–5793

Barton DHR, Werstiuk NH (1967) The constitution and stereochemistry of culmorin. Chem Commun:30–31. https://doi.org/10.1039/C19670000030

Bennett JW, Klich M (2003) Mycotoxins. Clin Microbiol Rev 16(3):497–516. https://doi.org/10.1128/CMR.16.3.497-516.2003

Bentley R (2000) Mycophenolic acid: a one hundred year odyssey from antibiotic to immunosuppressant. Chem Rev 100:3801–3826

Bhatnagar D, Ehrlich KC, Moore GG, Payne GA (2014) *Aspergillus-Aspergillus flavus*. In: Encyclopedia of food microbiology, vol 1. Elsevier, Amsterdam, pp 83–91

Bianchini A, Bullerman LB (2014) Mycotoxins. Classification. In: Batt CA, Tortorello ML (eds) Encyclopedia of food microbiology, 2nd edn. Academic Press, London, pp 854–861. https://doi.org/10.1016/B978-0-12-384730-0.00230-5. ISBN 9780123847331

Birkenshaw JH (1955) Ultraviolet absorption spectra of some polyhydroxyanthraquinones. Biochem J 59:485–486

Birkinshaw JH, Oxford AE, Raistrick H (1936) Studies in the biochemistry of microorganisms. XLVIII. Penicillic acid, a metabolic product of *Penicillium puberulum* Bainier and *Penicillium cyclopium* Westling. Biochem J 30:394–411

Blackwell BA, Miller JD, Greenhalgh R (1985) ^{13}C NMR study of the biosynthesis of toxins by *Fusarium graminearurn*. J Biol Chem 260(7):4243–4247

Bokhari FM, Flannigan B (1996) Synthesis of penicillic acid and other metabolites by *Penicillium aurantiogriseum* in stored cereals and lentils. In: Sand W (ed) Biodeterioration and biodegradation, Dachema Monographs 133. Dachema, Frankfurt, pp 237–244

Braun D, Ezekiel CN, Marko D, Warth B (2020) Exposure to Mycotoxin-mixtures via breast milk: an ultra-sensitive LC-MS/MS biomonitoring approach. Front Chem 8:423

Bryden W (2009) Mycotoxins and mycotoxicoses: significance, occurrence and mitigation in the food chain. In: General, applied and systems toxicology. Wiley, New York, NY, pp 3529–3553

Bullerman LB (2003) Mycotoxins - classifications. In: Encyclopedia of food sciences and nutrition. Elsevier, Amsterdam, pp 4080–4089. https://doi.org/10.1016/B0-12-227055-X/00821-X. ISBN 978-0-12-227055-0

Burmeister HR, Ciegler A, Vesonder RF (1979) Moniliformin, a metabolite of *Fusarium moniliforme* NRRL 6322: purification and toxicity. Appl Environ Microbiol 37(1):11–13. https://doi.org/10.1128/aem.37.1.11-13.1979

Callebaut F, Tangni EK, Debongnie P, Stals E, Huybrechts B, Waegeneers N, Delezie E, Van Pamel E, Daeseleire E (2011) Carry-over of mycotoxins to animal products: case study poultry. Sci Rep 211(212):2011–2012

Castoria R, Lima G, Ferracane R, Ritieni A (2005) Occurrence of mycotoxin in farro samples from Southern Italy. J Food Prot 68:416–420

Chan PK, Hayes AW, Siraj MY, Meydrech EF (1984) Pharmacokinetics of the mycotoxin penicillic acid in male mice: absorption, distribution, excretion, and kinetics. Toxicol Appl Pharmacol 73(2):195–203

Chang PK, Ehrlich KC, Fujii I (2009) Cyclopiazonic acid biosynthesis of *Aspergillus flavus* and *Aspergillus oryzae*. Toxins 1(2):74–99

Chang MH, Huang FJ, Chan WH (2012) Emodin induces embryonic toxicity in mouse blastocysts through apoptosis. Toxicology 299(1):25–32. https://doi.org/10.1016/j.tox.2012.05.006

Chaouloff F, Laude D, Merino D, Serrurrier B, Elghozi JL (1986) Peripheral and central short term effects of fusaric acid, a DBH inhibitor, on tryptophan and serotonin in the rat. Neur Transm 65: 219–232

Chen BH, Inbaraj BS (2016) Nanomaterial-based sensors for mycotoxin analysis in food. In: Grumezescu AM (ed) Nanotechnology in the agri-food industry, novel approaches of nanotechnology in food. Academic Press, London, pp 387–423. https://doi.org/10.1016/B978-0-12-804308-0.00012-1. ISBN 9780128043080

Chen S, Xu X, Dai X, Yang C, Qiang S (2007) Identification of tenuazonic acid as a novel type of natural photosystem II inhibitor binding in QB-site of Chlamydomonas reinhardtii. Biochim Biophys Acta Bioenerg 1767(4):306–318

Cholewinski G, Malachowska-Ugarte M, Dzierzbicka K (2010) The chemistry of mycophenolic acid – synthesis and modifications towards desired biological activity. Curr Med Chem 17: 1926–1941

Chu FS (2003) Mycotoxins. Toxicology. In: Encyclopedia of food sciences and nutrition, 2nd edn. Academic, London, pp 4096–4108

Chu Q, Wu W, Smalley EB (1993) Decreased cell-mediated immunity and lack of skeletal problems in broiler chickens consuming diets amended with fusaric acid. Avian Dis 37:863–867

Cole RJ, Cox RH (1981) Handbook of toxic fungal metabolites. Academic Press, New York, NY

Cole RJ, Kirksey JW, Cutler HG, Doupnik BL, Peckham JC (1973) Toxin from *Fusarium moniliforme*: effects on plants and animals. Science 179:1324–1326

Curry PT, Reed RN, Martino RM, Kitchin RM (1984) Induction of sister-chromatid exchanges *in-vivo* in mice by the mycotoxins sterigmatocystin and griseofulvin. Mutat Res 137(2–3):111–115

Dalefield R (2017) Mycotoxins and mushrooms. In: Dalefield R (ed) Veterinary toxicology for Australia and New Zealand. Elsevier, Amsterdam, pp 373–419. https://doi.org/10.1016/B978-0-12-420227-6.00020-7. ISBN 9780124202276

De Saeger S, Audenaert K, Croubels S (2016) Report from the 5th International Symposium on Mycotoxins and Toxigenic Moulds: Challenges and Perspectives (MYTOX) Held in Ghent, Belgium, May 2016. Toxins 8:146. Mycotoxin Contamination in Sugar Cane Grass and Juice:

First Report on Multi-Toxins Detection and Exposure Assessment in Humans. Abdallah, M.F.; Krska, R.; Sulyok, M

Decleer M, Landschoot S, De Saeger S, Rajkovic A, Audenaert K (2018) Impact of fungicides and weather on cyclodepsipeptide-producing *Fusarium* spp. and beauvericin and enniatin levels in wheat grains. J Sci Food Agric 99:253–262. https://doi.org/10.1002/jsfa.9167

Desjardins AE (2006) *Fusarium* mycotoxins – chemistry, genetics and biology. APS Press, College Park, MD, p 260

Desjardins AE, Maragos CM, Proctor RH (2006) Maize ear rot and moniliformin contamination by cryptic species of *Fusarium subglutinans*. J Agric Food Chem 54:7383–7390

Devaraja S, Girish KS, Santhosh MS, Hemshekhar M, Nayaka SC, Kemparaju K (2013) Fusaric acid, a mycotoxin, and its influence on blood coagulation and platelet functio. Blood Coagul Fibrinolysis 24(4):419–423. https://doi.org/10.1097/MBC.0b013e32835d548c

Devreese M, De Baere S, De Backer P, Croubels S (2013) Quantitative determination of the *Fusarium* mycotoxins beauvericin, enniatin A, A1, B and B1 in pig plasma using high performance liquid chromatography-tandem mass spectrometry. Talanta 106:212–219

Diaz Toro PCD, Arevalo FJ, Zon MA, Fernandez H (2015) Studies of the electrochemical behavior of moniliformin mycotoxin and its sensitive determination at pretreated glassy carbon electrodes in a nonaqueous medium. J Electroanal Chem 738:40–46. https://doi.org/10.1016/j.jelechem.2014.11.026

Dobson ADW (2017) Mycotoxins in cheese. In: McSweeney PLH, Fox PF, Cotter PD, Everett DW (eds) Cheese, 4th edn. Academic Press, London, pp 595–601. https://doi.org/10.1016/B978-0-12-417012-4.00023-5. ISBN 9780124170124

Dong X, Fu J, Yin X, Cao S, Li X, Lin L, Huyiligeqi NJ (2016) Emodin: a review of its pharmacology, toxicity and pharmacokinetics. Phytother Res 30:1207–1218

Dorner JW, Cole RJ, Erlington DJ, Suksupath S, McDowell GH, Bryden WL (1994) Cyclopiazonic acid residues in milk and eggs. J Agric Food Chem 42(7):1516–1518

Dowd PF (1989) Toxicity of naturallyoccurring levels of the *Penicillium* mycotoxins citrinin, ochratoxin A and penicillic acid to the corn earworm, *Heliothis zea*, and the fall armyworm, *Spodoptera frugiperda* (Lepidoptera: Noctuidae). Environ Entomol 18:24–29

Drugs.com (2021). https://www.drugs.com/sfx/mycophenolic-acid-side-effects.html

EFSA (2014) Scientific opinion on the risks to human and animal health related to the presence of beauvericin and enniatins in food and feed. EFSA J 12(8):3802. https://doi.org/10.2903/j.efsa.2014.3802

EFSA Panel on Contaminants in the Food Chain (CONTAM), Knutsen HK, Alexander J, Barregard L, Bignami M, Bruschweiler B, Edler L et al (2018) Risks to human and animal health related to the presence of moniliformin in food and feed. EFSA J 16(3):5082. https://doi.org/10.2903/j.efsa.2018.5082

EFSA Panel on Contaminants in the Food Chain (CONTAM), Schrenk D, Bodin L, Chipman JK, Del Mazo J, Grasl-Kraupp B, Hogstrand C, Hoogenboom LR, Leblanc JC, Nebbia CS, Nielsen E, Ntzani E, Petersen A, Sand S, Schwerdtle T, Vleminckx C, Wallace H, Alexander J, Cottrill B, Dusemund B, Mulder P, Arcella D, Baert K, Cascio C, Steinkellner H, Bignami M (2019) Scientific opinion on the risks for animal and human health related to the presence of quinolizidine alkaloids in feed and food, in particular in lupins and lupin-derived products. EFSA J 17(11):e05860

El-Bondkly AMA (2014) Sequence analysis of industrially important genes from *Trichoderma*. In: Gupta VK, Schmoll M, Herrera-Estrella A, Upadhyay RS, Druzhinina I, Tuohy MG (eds) Biotechnology and biology of *Trichoderma*. Elsevier, Amsterdam, pp 377–392. https://doi.org/10.1016/B978-0-444-59576-8.00028-X

El-Shanawany AA, Mostafa ME, Barakat A (2005) Fungal populations and mycotoxins in silage in Assiut and Sohag governorates in Egypt, with a special reference to characteristic *Aspergilli* toxins. Mycopathologia 159(2):281–289

EMBL-EBI: EMBL's European Bioinformatics Institute (n.d.). http://www.ebi.ac.uk/chebi/searchId.do?chebiId=CHEBI:3000

European Food Safety Authority (EFSA) (2011) Submission of scientific peer-reviewed open literature for the approval of pesticide active substances under Regulation (EC) No 1107/2009. EFSA J 9(2):2092

European Food Safety Authority (EFSA) (2017) Dietary reference values for nutrients-summary report. https://doi.org/10.2903/sp.efsa.2017.e15121. Approved: 4 December 2017

Fernández-Blanco C, Font G, Ruiz MJ (2016) Interaction effects of enniatin B, deoxinivalenol and alternariol in Caco-2 cells. Toxicol Lett 24:38–48

Firáková S, Šturdíková M, Múčková M (2007) Bioactive secondary metabolites produced by microorganisms associated with plants. Biologia 62:251–257. https://doi.org/10.2478/s11756-007-0044-1

Flajs D, Peraica M (2009) Toxicological properties of citrinin. Arhiv za Higijenu Rada i Toksikologiju 60(4):457–464. https://doi.org/10.2478/10004-1254-60-2009-1992

Fotso J, Leslie JF, Smith JS (2002) Production of beauvericin, moniliformin, fusaproliferin, and fumonisins b(1), b(2), and b(3) by fifteen ex-type strains of *Fusarium* species. Appl Environ Microbiol 68(10):5195–5197. https://doi.org/10.1128/AEM.68.10.5195-5197.2002

Fraeyman S, Devreese M, Antonissen G, De Baere S, Rychlik M, Croubels S (2016) Comparative oral bioavailability, toxicokinetics, and biotransformation of enniatin B1 and enniatin B in broiler chickens. J Agric Food Chem 64(38):7259–7264

Fremy JM, Alassane-Kpembi I, Oswald IP, Cottrill B, Van Egmond HP (2019) A review on combined effects of moniliformin and co-occurring *Fusarium* toxins in farm animals. World Mycotoxin J 12(3):281–291. https://doi.org/10.3920/WMJ2018.2405

Frisvad JC (2018) A critical review of producers of small lactone mycotoxins: patulin, penicillic acid and moniliformin. World Mycotoxin J 11(1):73–100. https://doi.org/10.3920/WMJ2017.2294

Frisvad JC, Larsen TO (2015) Chemodiversity in the genus *Aspergillus*. Appl Microbiol Biotechnol 99:7859–7877

Frisvad JC, Thrane U, Samson RA, Pitt JI (2006) Important mycotoxins and the fungi which produce them. Adv Exp Med Biol 571:3–31. https://doi.org/10.1007/0-387-28391-9_1. PMID: 16408591

Gammelsrud A, Solhaug A, Dendelé B, Sandberg WJ, Ivanova L, Kocbach Bølling A, Lagadic-Gossmann D, Refsnes M, Becher R, Eriksen G, Holme JA (2012) Enniatin B-induced cell death and inflammatory responses in RAW 267.4 murine macrophages. Toxicol Appl Pharmacol 261(1):74–87

Gao W, Jiang L, Ge L, Chen M, Geng C, Yang G, Li Q, Ji F, Yan Q, Zou Y, Zhong L, Liu X (2015) Sterigmatocystin-induced oxidative DNA damage in human liver-derived cell line through lysosomal damage. Toxicol in Vitro 29(1):1–7. https://doi.org/10.1016/j.tiv.2014.08.007

García-Herranz V, Valdehita A, Navas JM, Fernández-Cruz ML (2019) Cytotoxicity against fish and mammalian cell lines and endocrine activity of the mycotoxins beauvericin, deoxynivalenol and ochratoxin-A. Food Chem Toxicol 127:288–297

Ghadevaru G (2013) *In vitro* evaluation of toxin binding efficacy for aflatoxin B1, ochratoxin A, citrinin, zearalenone, T2 toxin, penicillic acid and fumonisin in brouiler chickens. J Vet Pharmacol Ther 35(3):50

Ghebremeskel M, Langseth W (2001) The occurrence of culmorin and hydroxy-culmorins in cereals. Mycopathologia 152(2):103–108. https://doi.org/10.1023/a:1012479823193

Gil-Serna J, Vázquez C, González-Jaén MT, Patiño B (2014a) Mycotoxins. toxicology. In: Encyclopedia of food microbiology, vol 2. Elsevier, Amsterdam, pp 887–892. https://doi.org/10.1016/B978-0-12-384730-0.00234-2

Goeger DE, Riley RT (1989) Interaction of cyclopiazonic acid with rat skeletal muscle sarcoplasmic reticulum vesicles. Effect on Ca2+ binding and Ca2+ permeability. Biochem Pharmacol 38(22):3995–4003

Gore T, Panse T, Venkataraman K (1946) Citrinin. Nature 157:333. https://doi.org/10.1038/157333a0

Gqaleni N, Smith JE, Lacey J, Gettinby G (1997) Effects of temperature, water activity, and incubation time on production of aflatoxins and cyclopiazonic acid by an isolate of *Aspergillus flavus* in surface agar culture. Appl Environ Microbiol 63(3):1048–1053

Granby K, Mortensen A, Broesboel-Jensen B (2012) Potential contamination issues arising from the use of biofuel and food industry by-products in animal feed. In: Fink-Gremmels J (ed) Woodhead Publishing series in food science, Technology and nutrition, animal feed contamination. Woodhead Publishing, London, pp 514–539. https://doi.org/10.1533/9780857093615.5.514

Griffin GF, Chu FS (1983) Toxicity of the *Alternaria* metabolites alternariol, alternariol methyl ether, altenuene, and tenuazonic acid in the chicken embryo assay. Appl Environ Microbiol 46(6):1420–1422

Groopman JD, Kensler TW, Wu F (2013) Food safety: mycotoxins–occurrence and toxic effects. In: Caballero B (ed) Encyclopedia of human nutrition, 3rd edn. Academic Press, London, pp 337–341. https://doi.org/10.1016/B978-0-12-375083-9.00122-7

Gross M, Levy R, Toepke H (1984) Occurrence and analysis of the mycotoxin emodin. Nahrung 28(1):31–44. https://doi.org/10.1002/food.19840280116

Gruber-Dorninger C, Novak B, Nagl V, Berthiller F (2017) Emerging mycotoxins: beyond traditionally determined food contaminants. J Agric Food Chem 65(33):7052–7070. https://doi.org/10.1021/acs.jafc.6b03413

Habler K, Gotthardt M, Schüler J, Rychlik M (2017) Multi-mycotoxin stable isotope dilution LC-MS/MS method for Fusarium toxins in beer. Food Chem 218:447–454

Hai Y, Chen M, Huang A, Tang Y (2020) Biosynthesis of mycotoxin fusaric acid and application of a PLP-dependent enzyme for chemoenzymatic synthesis of substituted l-pipecolic acids. J Am Chem Soc 142(46):19668–19677. https://doi.org/10.1021/jacs.0c09352

Hallas-Moeller M, Nielsen KF, Frisvad JC (2016) Production of the *Fusarium* mycotoxin moniliformin by *Penicillium melanoconidium*. J Agric Food Chem 64:4505–4510

Hasan HA (1998) Studies on toxigenic fungi in roasted foodstuff (salted seed) and halotolerant activity of emodin-producing *Aspergillus wentii*. Folia Microbiol (Praha) 43(4):383–391. https://doi.org/10.1007/BF02818578

Hashizume H, Nishimura Y (2008) Cyclic lipopeptide antibiotics. In: Atta-ur R (ed) Studies in natural products chemistry, vol 35. Elsevier, Amsterdam, pp 693–751. https://doi.org/10.1016/S1572-5995(08)80016-6

He Q, Liu K, Wang S, Hou H, Yuan Y, Wang X (2012) Toxicity induced by emodin on zebrafish embryos. Drug Chem Toxicol 36:149–154

Hetherington AC, Raistrick H (1931) Studies in the biochemistry of micro-organisms. Part XIV: On the production and chemical constitution of a new yellow colouring matter, citrinin, produced from glucose by Penicillium. Phil Transa R Soc Lond B Biol Sci 220(468–473):269–295

Hidaka H, Nagatsu T, Takeya K (1969) Fusaric acid, a hypotensive agent produced by fungi. J Antibiot 22:228–230

Hirota M, Menta A, Yoneyama K, Kitabatake N (2002) A major decomposition product, citrinin h2, from citrinin on heating with moisture. Biosci Biotechnol Biochem 66(1):206–210. https://doi.org/10.1271/bbb.66.206

Hoque N, Hasan CM, Rana MS, Varsha A, Sohrab MH, Rahman KM (2018) Fusaproliferin, a fungal mycotmoxin, shows cytotoxicity against pancreatic cancer cell lines. Molecules 23(12):3288. https://doi.org/10.3390/molecules23123288

Hu L, Rychlik M (2014) Occurrence of enniatins and beauvericin in 60 Chinese medicinal herbs. Food Additiv Contam A 31:1240–1245. https://doi.org/10.1080/19440049.2014.913813

Huang S, Wang J, Xing L, Shen H, Yan X, Wang J, Zhang X (2014) Impairment of cell cycle progression by sterigmatocystin in human pulmonary cells *in vitro*. Food Chem Toxicol 66:89–95

Huiqin W, Shufen Z, Taifeng L, Dawei Z (2017) Introduction and safety evaluation of citrinin in foods. J Food Nutr Sci 5(5):179–183. https://doi.org/10.11648/j.jfns.20170505.13

International Agency for Research on Cancer (IARC) (2014) Monographs on the Evaluation of carcinogenic risks to humans. IARC, Lyon

Ismaiel AA, Papenbrock J (2015) Mycotoxins: producing fungi and mechanisms of phytotoxicity. Agriculture 5:492–537

Iwahashi H, Kitagawa E, Suzuki Y, Ueda Y, Ishizawa Y-h, Nobumasa H, Kuboki Y, Hosoda H, Iwahashi Y (2007) Evaluation of toxicity of the mycotoxin citrinin using yeast ORF DNA microarray and Oligo DNA microarray. BMC Genomics 8:95. https://doi.org/10.1186/1471-2164-8-95

Iwasaki S, Muro H, Nozoe S, Okuda S, Sato Z (1972) Isolation of 3, 4-dihydro-3, 4, 8-trihydroxy-1 (2H)-naphthalenone and tenuazonic acid from *Pyricularia oryzae cavara*. Tetrahedron Lett 13: 13–16

Jajić I, Dudaš T, Krstović S, Krska R, Sulyok M, Bagi F, Savić Z, Guljaš D, Stankov A (2019) Emerging *Fusarium* mycotoxins fusaproliferin, beauvericin, enniatins, and moniliformin in Serbian maize. *Toxins* 11(6):357. https://doi.org/10.3390/toxins11060357

Jakšić DD, Kocsubé S, Bencsik O, Kecskeméti A, Szekeres A, Vágvölgyi C, Varga J, Šegvić KM (2017) New sterigmatocystin-producing species of *Aspergillus* section *Versicolores* from indoor air in Croatia. Mycol Prog 16(1):63–72

Javed T, Bunte RM, Dombrink-Kurtzman MA, Richard JL, Bennett GA, Cote LM, Buck WB (2005) Comparative pathologic changes in broiler chicks on feed amended with *Fusarium proliferatum* culture material or purified fumonisin B1 and moniliformin. Mycopathologia 159:553–564

Jestoi M (2008) Emerging *Fusarium*-mycotoxins fusaproliferin, beauvericin, enniatins, and moniliformin: a review. Crit Rev Food Sci Nutr 48(1):21–49. https://doi.org/10.1080/10408390601062021

Jestoi M, Ritieni A, Rizzo A (2004) Analysis of the *Fusarium*-mycotoxins fusaproliferin and trichothecenes in grains using gas chromatography-mass spectrometry. J Agric Food Chem 52:1464–1469

Johnson JR, Bruce WF, Dutcher JD (1943) Gliotoxin, The Antibiotic Principle of *Gliocladium fimbriatum*. I. Production, physical and biological properties1. J Am Chem Soc 65(10): 2005–2009. https://doi.org/10.1021/ja01250a051

Juan C, Ritieni A, Mañes J (2013) Occurrence of *Fusarium mycotoxins* in Italian cereal and cereal products from organic farming. Food Chem 141:1747–1755. https://doi.org/10.1016/j.foodchem.2013.04.061

Kalayou S, Ndossi D, Frizzell C, Groseth PK, Connolly L, Sørlie M, Verhaegen S, Ropstad E (2015) An investigation of the endocrine disrupting potential of enniatin B using in vitro bioassays. Toxicol Lett 233(2):84–94

Kataoka T, Nagai K (2002) Molecular dissection of cytotoxic functions mediated by T cells. In: Endo I, Kudo T, Osada H, Shibata T, Yamaguchi I (eds) Progress in biotechnology, vol 22. Elsevier, Amsterdam, pp 13–23. https://doi.org/10.1016/S0921-0423(02)80039-9

Keblys M, Bernhoft A, Höfer CC, Morrison E, Jørgen H, Larsen S, Flåøyen A (2004) The effects of the *Penicillium* mycotoxins citrinin, cyclopiazonic acid, ochratoxinA, patulin, penicillic acid, and roquefortine C on *in vitro* proliferation of porcine lymphocytes. Mycopathologia 158:317–324

Knowles SL, Mead ME, Silva LP, Raja HA, Steenwyk JL, Goldman GH, Oberlies NH, Rokas A, Bahn Y-S, Doyle S, Bok JW (2020) Gliotoxin, a known virulence factor in the major human pathogen *Aspergillus fumigatus*, is also biosynthesized by its nonpathogenic relative *Aspergillus fischeri*. MBio 11(1):e03361–e03319. https://doi.org/10.1128/mBio.03361-19

Kögl F, Postowsky JJ (1925) Untersuchungen fiber Pilzfarbstoffe. II. Uber die Farbstaffe des blutroten Hautkorpfes (*Dermocybe sanquinea* Wulf.). Justus Liebigs Annalen der Chemie 444:1–7

Kosuri NR, Grove MD, Yates SG, Tallent WH, Ellis JJ, Wolff IA, Nichols RE (1970) Response of cattle to mycotoxins of *Fusarium tricinctum* isolated from corn and fescue. J Am Vet Med Assoc 157:938–940

Kovalenko AV, Soldatenko NA, Fetisov LN, Strel'tsov NV (2011) More accurate determination of the minimum allowable level of sterigmatocystin in piglet feed. Russian Agricultural. Science 37(6):504–507

Křížová L, Dadáková K, Dvořáčková M, Kašparovský T (2021) Feedborne mycotoxins beauvericin and enniatins and livestock animals. Toxins (Basel) 13(1):32. https://doi.org/10.3390/toxins13010032

Krska R, Nährer K, Richard JL, Rodrigues I, Schuhmacher R, Slate AB, Whitaker TB (2012) Guide to mycotoxins featuring mycotoxin risk management in animal production. BIOMIN, Herzogenburg

Kumari A, Singh K (2021) Evaluation of prophylactic efficacy of cinnamaldehyde in murine model against *Paradendryphiella arenariae* mycotoxin tenuazonic acid-induced oxidative stress and organ toxicity. Sci Rep 11(1):1–16

Kumari A, Tirkey NN (2019) Tenuazonic acid: a potent mycotoxin. In: Singh K, Shrivastava N (eds) Recent trends in human and animal mycology. Springer Nature, Berlin, pp 203–211. ISBN: 978-981-13-9435-5

Kwon-Chung KJ, Sugui JA (2009) What do we know about the role of gliotoxin in the pathobiology of *Aspergillus fumigatus*? Med Mycol 47(Suppl 1):S97–S103. https://doi.org/10.1080/13693780802056012

Langseth W, Ghebremeskel M, Kosiak B, Kolsaker P, Miller D (2001) Production of culmorin compounds and other secondary metabolites by *Fusarium culmorum* and *F. graminearum* strains isolated from Norwegian cereals. Mycopathologia 152(1):23–34. https://doi.org/10.1023/a:1011964306510

Leslie JF, Zeller KA, Lamprecht SC, Rheeder JP, Marasas WFO (2005) Toxicity, pathogenicity, and genetic differentiation of five species of *Fusarium* from sorghum and millet. Phytopathology 95:275–283

Lhotská I, Šatínský D, Havlíková L, Solich P (2016) A fully automated and fast method using direct sample injection combined with fused-core column on-line SPE-HPLC for determination of ochratoxin A and citrinin in lager beers. Anal Bioanal Chem 408(12):3319–3329. https://doi.org/10.1007/s00216-016-9402-6

Li Y, Zhou YC, Yang MH, Ou-Yang Z (2012) Natural occurrence of citrinin in widely consumed traditional Chinese food red yeast rice, medicinal plants and their related products. Food Chem 132(2):1040–1045. https://doi.org/10.1016/j.foodchem.2011.11.051

Liu X, Walsh CT (2009) Cyclopiazonic acid biosynthesis in *Aspergillus* sp.: characterization of a reductase-like R* domain in cyclopiazonate synthetase that forms and releases cyclo-acetoacetyl-L-tryptophan. Biochemistry 48(36):8746–8757. https://doi.org/10.1021/bi901123

Liu Y, Jia L, Liu ZC, Zhang H, Zhang PJ, Wan Q, Wang R (2009) Emodin ameliorated high-glucose induced mesangial P38 over-activation and hypocontractility via activating PPARγ. Exp Mol Med 41:648–655

Logrieco A, Moretti A, Fornelli F, Fogliano V, Ritieni A, Caiaffa MF, Randazzo G, Bottalico A, Macchia L (1996) Fusaproliferin production by *Fusarium subglutinans* and its toxicity to *Artemia salina*, SF-9 insect cells, and IARC/LCL 171 human B lymphocytes. Appl Environ Microbiol 62:3378–3384

Logrieco A, Rizzo A, Ferracane R, Ritien A (2002) Occurrence of beauvericin and enniatins in wheat affected by *Fusarium avenaceum* head blight. Appl Environ Microbiol 68(1):82–85. https://doi.org/10.1128/AEM.68.1.82-85.2002

Ludueña RF, Prasad V, Roach MC, Lacey E (1989) The interaction of phomopsin A with bovine brain tubulin. Arch Biochem Biophys 272(1):32–38. https://doi.org/10.1016/0003-9861(89)90191-4

Lurie A, Katz J, Ludwin SK, Seftel HC, Metz J (1969) Platelet life-span and sites of platelet sequestration in onyalai. Br Med J 4(5676):146–148. https://doi.org/10.1136/bmj.4.5676.146

Luz C, Saladino F, Luciano FB, Mañes J, Meca G (2017) Occurrence, toxicity, bioaccessibility and mitigation strategies of beauvericin, a minor *Fusarium* mycotoxin. Food Chem Toxicol 107:430–439. https://doi.org/10.1016/j.fct.2017.07.032

Mahnine N, Meca G, Elabidi A, Fekhaoui M, Saoiabi A, Font G, Mañes J, Zinedine A (2011) Further data on the levels of emerging *Fusarium* mycotoxins enniatins (A, A1, B, B1), beauvericin and fusaproliferin in breakfast and infant cereals from Morocco. Food Chem 124: 481–485. https://doi.org/10.1016/j.foodchem.2010.06.058

Manizan AL, Oplatowska-Stachowiak M, Piro-Metayer I, Campbell K, Koffi-Nevry R, Elliott C, Akaki D, Montetc D, Brabet C (2018) Multi-mycotoxin determination in rice, maize and peanut products most consumed in Cote d'Ivoire by UHPLC-MS/MS. Food Control 87:22–30. https://doi.org/10.1016/j.foodcont.2017.11.032

Maranghi F, Tassinari R, Narciso L, Tait S, Rocca CL, Felice GD, Butteroni C, Corinti S, Barletta B, Cordelli E (2018) *In vivo* toxicity and genotoxicity of beauvericin and enniatins. Combined approach to study *in vivo* toxicity and genotoxicity of mycotoxins beauvericin (BEA) and enniatin B (ENNB). EFSA Support Publ 15:1406E

Markov K, Pleadin J, Bevardi M, Vahčić N, Sokolić-Mihalak D, Frece J (2013) Natural occurrence of aflatoxin B1, ochratoxin A and citrinin in Croatian fermented meat products. Food Control 34:312–317

Matthees D (1983) Determination of emodin in feeds. J Agric Food Chem 31(2):453–454. https://doi.org/10.1021/jf00116a070

Matxalen V-G, Sergio R, Pilar DM, Patricia M, Cristina C, Meis Jacques F, Antonio R, Julian P, Galvez Eva M (2018) Production of the invasive aspergillosis biomarker bis(methylthio)-gliotoxin within the genus *Aspergillus*: *in vitro* and *in vivo* metabolite quantification and genomic analysis. Frontiers in Microbiology 9:1246. https://doi.org/10.3389/fmicb.2018.01246

Meca G, Zinedine A, Blesa J, Font G, Mañes J (2010) Further data on the presence of *Fusarium* emerging mycotoxins enniatins, fusaproliferin and beauvericin in cereals available on the Spanish markets. Food Chem Toxicol 48(5):1412–1416. https://doi.org/10.1016/j.fct.2010.03.010

Meronuck RA, Steele JA, Mirocha CJ, Christensen CM (1972) Tenuazonic acid, a toxic produced by *Alternaria alternata*. Appl Microbiol 23(3):613–617. https://doi.org/10.1128/am.23.3.613-617.1972

Micco C, Miraglia M, Onori R, Libanori A, Brera C, Mantovani A, Macri C (1991) Effect of combined exposure to ochratoxin A and penicillic acid on residues and toxicity in broilers. Int J Food Sci Nutr 20:101–108

Miller JD, MacKenzie S (2000) Secondary metabolites of *Fusarium venenatum* strains with deletions in the Tri5 gene encoding trichodiene synthetase. Mycologia 92:764–771

Miller JD, Sun M, Gilyan A, Roy J, Rand TG (2010) Inflammation-associated gene transcription and expression in mouse lungs induced by low molecular weight compounds from fungi from the built environment. Chem Biol Interact 183(1):113–124

Moretti A, Mulè G, Ritieni A, Logrieco A (2007) Further data on the production of beauvericin, enniatins and fusaproliferin and toxicity to *Artemia salina* by *Fusarium* species of *Gibberella fujikuroi* species complex. Int J Food Microbiol 118:158–163

Mori H, Sugie S, Yoshimi N, Kitamura J, Niwa M, Hamasaki T, Kawai K (1986) Genotoxic effects of a variety of sterigmatocystin-related compounds in the hepatocyte/DNA-repair test and the *Salmonella* microsome assay. Mutat Res 173(3):217–222

Mueller SO, Eckert I, Lutz WK, Stopper H (1996) Genotoxicity of the laxative drug components emodin, aloe-emodin and danthron in mammalian cells: topoisomerase II mediated? Mutat Res 371:165–173

Munkvold G, Stahr HM, Logrieco A, Moretti A, Ritieni A (1998) Occurrence of fusaproliferin and beauvericin in *Fusarium*-contaminated livestock feed in Iowa. Appl Environ Microbiol 64(10): 3923–3926. https://journals.asm.org/doi/abs/10.1128/AEM.64.10.3923-3926.1998

Nagatsu T, Hidaka H, Kuzuya J, Takeya K, Umezawa H, Takeuchi T, Suda H (1970) Inhibition of dopamine beta-hydroxylase by fusaric acid (5-butylpicolinic acid) *in vitro* and *in vivo*. Biochem Pharmacol 19:35–44

Natori S, Sato F, Udagawa S (1965) Anthraquinone metabolites of *Talaromyces avellanens* (Thomet Turreson), C. R. Benjamin and *Preussia multispora* (Saito et Minoura) Cain. Chem Pharm Bull 13:385–389

Nieto CHD, Granero AM, Zon MA, Fernández H (2018) Sterigmatocystin: a mycotoxin to be seriously considered. Food Chem Toxicol 118:460–470. https://doi.org/10.1016/j.fct.2018.05.057

Nirwan S, Kakkar R (2019) Rhinovirus RNA polymerase: structure, function, and inhibitors. In: Gupta SP (ed) Viral polymerases. Academic Press, London, pp 301–331. https://doi.org/10.1016/B978-0-12-815422-9.00011-5

Nishie K, Cole RJ, Dorner JW (1985) Toxicity and neuropharmacology of cyclopiazonic acid. Food Chem Toxicol 23(9):831–839. https://doi.org/10.1016/0278-6915(85)90284-4

Oh SY, Cedergreen N, Yiannikouris A, Swamy HVLN, Karrow NA (2015) Assessing interactions of binary mixtures of *Penicillium* mycotoxins (PMs) by using a bovine macrophage cell line (BoMacs). Toxicol Appl Pharmacol 318:33–40

de Oliveira Filho JWG, Islam MT, Ali ES, Uddin SJ, Santos JVO, de Alencar MVOB, Júnior ALG, Paz MFCJ, de Brito MDRM, Sousa JMCE, Shaw S, de Medeiros MDGF, Dantas SMMM, Rolim HML, Ferreira PMP, Kamal MA, Pieczynska MD, Das N, Gupta VK, Mocan A, Dos Santos Andrade TJA, Singh BN, Mishra SK, Atanasov AG, Melo-Cavalcante AAC (2017) A comprehensive review on biological properties of citrinin. Food Chem Toxicol 110:130–141. https://doi.org/10.1016/j.fct.2017.10.002

Gil-Serna J, Vázquez C, González-Jaén MT, Patiño B (2014b) Mycotoxins. toxicology. In: Encyclopedia of food microbiology, vol 2. Elsevier, Amsterdam, pp 887–892. https://doi.org/10.1016/B978-0-12-384730-0.00234-2

Oliveira CA, Rosmaninho J, Rosim R (2006) Aflatoxin M1 and cyclopiazonic acid in fluid milk traded in São Paulo, Brazil. Food Addit Contam 23(2):196–201. https://doi.org/10.1080/02652030500398379

Ostry V (2008) *Alternaria* mycotoxins: an overview of chemical characterization, producers, toxicity, analysis and occurrence in foodstuffs. World Mycotoxin J 1(2):175–188. https://doi.org/10.3920/WMJ2008.x013

Ostry V, Malir F, Ruprich J (2013) Producers and important dietary sources of ochratoxin A and citrinin. Toxins (Basel) 5(9):1574–1586. https://doi.org/10.3390/toxins5091574

Ostry V, Toman J, Grosse Y, Malir F (2018) Cyclopiazonic acid: 50th anniversary of its discovery. World Mycotoxin J 11(1):135–148

Pandiyan V, Nayeem M, Nanjappan K, Ramamurti R (1990) Penicillic acid as Na^+, K^+ and Ca^{2+} channel blocker in isolated frog heart at toxic levels. Indian J Exp Biol 28:295–296

Pedersen PB, Miller JD (1999) The fungal metabolite culmorin and related compounds. Nat Toxins 7:305–309

Piecková E (2015) Domestic environment – indoor mycobiota as a public health risk factor. In: Viegas C, Pinheiro C, Sabino R, Viegas S, Brandao J, Veríssimo C (eds) Environmental mycology in public health. Fungi and mycotoxins risk assessment and management. Elsevier, Academic Press, London, pp 129–146. https://www.elsevier.com/books/environmental-mycology-in-public-health/viegas/978-0-12-411471-5

Pirrung MC, Nauhaus SK, Singh B (1996) Cofactor-directed, time-dependent inhibition of thiamine enzymes by the fungal toxin moniliformin. J Org Chem 61:2592–2593

Porter JK, Bacon CW, Wray EM, Hagler WM Jr (1995) Fusaric acid in *Fusarium moniliforme* cultures, corn, and feeds toxic to livestock and the neurochemical effects in the brain and pineal gland of rats. Nat Toxins 3:91–100

Prelusky D, Hamilton R, Trenholm H (1989) Application of the chick embryotoxicity bioassay for the evaluation of mycotoxin toxicity. Microbiol Alim Nutr 7:57–65

Prosperini A, Berrada H, Ruiz MJ, Caloni F, Coccini T, Spicer LJ, Perego MC, Lafranconi A (2017) A review of the mycotoxin enniatin B. Front Public Health 5:304. https://doi.org/10.3389/fpubh.2017.00304

PubChem (2019). https://pubchem.ncbi.nlm.nih.gov/compound/Fusaproliferin

PubChem (n.d.-a). https://pubchem.ncbi.nlm.nih.gov/compound/3007984
PubChem (n.d.-b). https://pubchem.ncbi.nlm.nih.gov/compound/alpha-Cyclopiazonic-acid
PubChem (n.d.-c). https://pubchem.ncbi.nlm.nih.gov/compound/Butenolides
PubChem (n.d.-d). https://pubchem.ncbi.nlm.nih.gov/compound/Citrinin
PubChem (n.d.-e). https://pubchem.ncbi.nlm.nih.gov/compound/Culmorin
PubChem (n.d.-f). https://pubchem.ncbi.nlm.nih.gov/compound/Emodin
PubChem (n.d.-g). https://pubchem.ncbi.nlm.nih.gov/compound/Enniatin-B
PubChem (n.d.-h). https://pubchem.ncbi.nlm.nih.gov/compound/Fusaric-acid
PubChem (n.d.-i). https://pubchem.ncbi.nlm.nih.gov/compound/Gliotoxin
PubChem (n.d.-j). https://pubchem.ncbi.nlm.nih.gov/compound/Moniliformin
PubChem (n.d.-k). https://pubchem.ncbi.nlm.nih.gov/compound/Mycophenolic-acid
PubChem (n.d.-l). https://pubchem.ncbi.nlm.nih.gov/compound/Penicillic-acid
PubChem (n.d.-m). https://pubchem.ncbi.nlm.nih.gov/compound/Phomopsin-A
PubChem (n.d.-n). https://pubchem.ncbi.nlm.nih.gov/compound/Sterigmatocystin

Randazzo G, Foglianoa V, Ritieni A, Rossin LME, Scarallo A, Segred AL (1993) Proliferin, a new sesterterpene from *Fusarium proliferatum*. Tetrahedron 49:10883–10896. https://doi.org/10.1016/S0040-4020(01)80241-6

Ritieni A, Fogliano V, Randazzo G, Scarallo A, Logrieco A, Moretti A, Mannina L, Bottalico A (1995) Isolation and characterization of fusaproliferin, a new toxic metabolite from *Fusarium proliferatum*. Nat Toxins 3(1):17–20

Ritieni A, Monti SM, Randazzo G, Logrieco A, Moretti A, Peluso G, Ferracane R, Fogliano V (1997) Teratogenic effects of fusaproliferin on chicken. J Agric Food Chem 45:3039–3043

Rodríguez-Carrasco Y, Heilos D, Richter L, Süssmuth RD, Heffeter P, Sulyok M, Kenner L, Berger W, Dornetshuber-Fleiss R (2016) Mouse tissue distribution and persistence of the food-born fusariotoxins enniatin B and beauvericin. Toxicol Lett 247:35–44

RSC (2018). https://www.rsc.org/MerckIndex/monograph/m10562/tenuazonic%20acid?q=authorize

Santini A, Ritieni A, Fogliano V, Randazzo G, Mannina L, Logrieco A, Benedetti E (1996) Structure and absolute stereochemistry of fusaproliferin, a toxic metabolite from *Fusarium proliferatum*. J Nat Prod 59(2):109–112. https://doi.org/10.1021/np960023k

Santini A, Meca G, Uhlig S, Ritieni A (2012) Fusaproliferin, beauvericin and enniatins: occurrence in food—a review. World Mycotoxin J 5(1):71–81. https://doi.org/10.3920/WMJ2011.1331

Scarpino V, Blandino M, Negre M, Reyneri A, Vanara F (2013) Moniliformin analysis in maize samples from North-West Italy using multifunctional clean-up columns and the LS-MS/MS detection method. Food Addit Contam A 30:876–884. https://doi.org/10.1080/19440049.2013.793825

Scauflaire J, Gourgue M, Callebaut A, Munaut F (2012) *Fusarium temperatum*, a mycotoxin-producing pathogen of maize. Eur J Plant Pathol 133:911–922

Scharf DH, Brakhage AA, Mukherjee PK (2016) Gliotoxin--bane or boon? Environ Microbiol 18(4):1096–1109. https://doi.org/10.1111/1462-2920.13080

Schenzel J, Hungerbühler K, Bucheli TD (2012) Mycotoxins in the environment: II. Occurrence and origin in Swiss river waters. Environ Sci Technol 46:13076–13084. https://doi.org/10.1021/es301558v

Schobert R, Schlenk A (2008) Tetramic and tetronic acids: an update on new derivatives and biological aspects. Bioorg Med Chem 16(8):4203–4221. https://doi.org/10.1016/j.bmc.2008.02.069

Selim ME, El-Gammal NA (2015) Role of fusaric acid mycotoxin in pathogensis process of tomato wilt disease caused by Fusarium oxysporum. J Bioprocess Biotech 5(10):1

Sharma D, Asrani RK, Ledoux DR, Rottinghaus GE, Gupta VK (2012) Toxic interaction between fumonisin B1 and moniliformin for cardiac lesions in Japanese quail. Avian Dis 56:545–554

Shephard GS, Thiel PG, Sydenham EW, Vleggaar R, Marasas WFO (1991) Reversed-phase high-performance liquid chromatography of tenuazonic acid and related tetramic acids. J Chromatogr B Biomed Sci Appl 66(1):195–205. https://doi.org/10.1016/0378-4347(91)80124-U

Shephard GS, Sewram V, Nieuwoudt TW, Marasas WFO, Ritieni A (1999) Production of the mycotoxins fusaproliferin and beauvericin by South African isolates in the *Fusarium* section *liseola*. J Agric Food Chem 47:5111–5115

Shi Y, Yang W, Tang X, Yan Q, Cai X, Wu F (2021) Keshan disease: a potentially fatal endemic cardiomyopathy in remote mountains of China. Front Pediatr 9:576916. https://doi.org/10.3389/fped.2021.576916

Shigeura HT (1967) Tenuazonic acid. In: Antibiotics. Springer, Berlin, pp 360–365

Shigeura HT, Gordon CN (1963) The biological activity of tenuazonic acid. Biochemistry 2(5): 1132–1137

Sifou A, Meca G, Serrano AB, Mahnine N, El Abidi A, Mañes J, El Azzouzi M, Zinedine A (2011) First report on the presence of emerging *Fusarium* mycotoxins enniatins (A, A1, B, B1), beauvericin and fusaproliferin in rice on the Moroccan retail markets. Food Control 22:1826–1830

Silva LJG, Pereira AMPT, Pena A, Lino CM (2020) Citrinin in foods and supplements: a review of occurrence and analytical methodologies. Foods 10(1):14. https://doi.org/10.3390/foods10010014

Simonetti E, Roberts IN, Montecchia MS, Gutierrez-Boem FH, Gomez FM, Ruiz JA (2018) A novel *Burkholderia ambifaria* strain able to degrade the mycotoxin fusaric acid and to inhibit *Fusarium* spp. growth. Microbiol Res 206:50–59. https://doi.org/10.1016/j.micres.2017.09.008

Sláma J, Košler J, Condon DJ, Crowley JL, Gerdes A, Hanchar JM, Horstwood MSA, Morris GA, Nasdala L, Norberg N, Schaltegger U, Schoene B, Tubrett MN, Whitehouse MJ (2008) Plešovice zircon—a new natural reference material for U–Pb and Hf isotopic microanalysis. Chem Geol 249(1–2):1–35

Smith TK, MacDonald EJ (1991) Effect of fusaric acid on brain regional neurochemistry and vomiting behavior in swine. J Anim Sci 69:2044–2049

Smith TK, Sousadias MG (1993) Fusaric acid content of swine feedstuffs. J Agric Food Chem 41: 2296–2298

Sørensen LM, Nielsen KF, Jacobsen T, Koch AG, Nielsen PV, Frisvad JC (2008) Determination of mycophenolic acid in meat products using mixed mode reversed phase-anion exchange clean-up and liquid chromatography-high-resolution mass spectrometry. J Chromatogr A 1205(1–2): 103–108. https://doi.org/10.1016/j.chroma.2008.08.019

Sørensen LM, Mogensen J, Nielsen KF (2010) Simultaneous determination of ochratoxin A, mycophenolic acid and fumonisin B(2) in meat products. Anal Bioanal Chem 398(3): 1535–1542. https://doi.org/10.1007/s00216-010-4059-z

Sorenson WG, Simpson J (1986) Toxicity of penicillic acid for rat alveolar macrophages *in vitro*. Environ Res 41:505–513

Sougiannis AT, Enos RT, VanderVeen BN, Velazquez KT, Kelly B, McDonald S, Cotham W, Chatzistamou I, Nagarkatti M, Fan D, Murphy EA (2021) Safety of natural anthraquinone emodin: an assessment in mice. BMC Pharmacol Toxicol 22(1):9. https://doi.org/10.1186/s40360-021-00474-1

Springer JP, Clardy J, Cole RJ, Kirksey JW, Hill RK, Carlson RM, Isido JL (1974) Structure and synthesis of moniliformin, a novel cyclobutane microbial toxin. J Am Chem Soc 96:2267–2268

Stanciu O, Juan C, Miere D, Loghin F, Mañes J (2017) Presence of enniatins and beauvericin in romanian wheat samples: from raw material to products for direct human consumption. Toxins 9:E189. https://doi.org/10.3390/toxins9060189

Steyn PS, Rabie CJ (1976) Characterization of magnesium and calcium tenuazonate from *Phoma sorghina*. Phytochemistry 15(12):1977–1979

Stoev SD (2017) Balkan endemic nephropathy–still continuing enigma, risk assessment and underestimated hazard of joint mycotoxin exposure of animals or humans. Chem Biol Interact 261:63–79. https://doi.org/10.1016/j.cbi.2016.11.018

Streit E, Schwab C, Sulyok M, Naehrer K, Krska R, Schatzmayr G (2013) Multi-mycotoxin screening reveals the occurrence of 139 different secondary metabolites in feed and feed ingredients. Toxins 5:504–523

Strongman D, Miller J, Calhoun L, Findlay J, Whitney N (1987) The biochemical basis for interference competition among some lignicolous marine fungi. Bot Mar 30:21–26

Svingen T, Lund Hansen N, Taxvig C, Vinggaard AM, Jensen U, Have RP (2017) Enniatin B and beauvericin are common in Danish cereals and show high hepatotoxicity on a high-content imaging platform. Environ Toxicol 32:1658–1664. https://doi.org/10.1002/tox.22367

Sweeney MJ, Gerzon K, Harris PN, Holmes RE, Poore GA, Williams RH (1972) Experimental antitumor activity and preclinical toxicology of mycophenolic acid. Cancer Res 32:1795–1802

Tabata S (2002) Mycotoxins. Aflatoxins and related compounds. In: Roginski H (ed) Encyclopedia of dairy sciences. Elsevier, Amsterdam, pp 2087–2095. https://doi.org/10.1016/B0-12-227235-8/00351-5. ISBN 9780122272356

Tabata S (2011) Yeasts and molds. Mycotoxins: aflatoxins and related compounds. In: Fuquay JW (ed) Encyclopedia of dairy sciences, 2nd edn. Academic Press, London, pp 801–811. https://doi.org/10.1016/B978-0-12-374407-4.00369-1. ISBN 9780123744074

Tabata S (2022) Mycotoxins: aflatoxins and related compounds. In: McSweeney PLH, McNamara JP (eds) Encyclopedia of dairy sciences, 3rd edn. Academic Press, London, pp 575–585. https://doi.org/10.1016/B978-0-12-818766-1.00071-4. ISBN 9780128187678

Tachibana M, Matsui C, Takeuchi Y, Suzuki E, Umezawa K (2008) Inhibition of NF-kappa B activation by penicillic acid and dihydropenicillic acid isolated from fungi. Heterocycles 76: 1561–1569

Taevernier L, Veryser L, Roche N, Peremans K, Burvenich C, Delesalle C, De Spiegeleer B (2016) Human skin permeation of emerging mycotoxins (beauvericin and enniatins). J Expos Sci Enviro Epidemiol 26:277–287. https://doi.org/10.1038/jes.2015.10

Tonshin AA, Teplova VV, Andersson MA, Salkinoja-Salonen MS (2010) The *Fusarium* mycotoxins enniatins and beauvericin cause mitochondrial dysfunction by affecting the mitochondrial volume regulation, oxidative phosphorylation and ion homeostasis. Toxicology 276(1): 49–57

Tookey HL, Yates SG, Ellis JJ, Grove MD, Nichols RE (1972) Toxic effects of a butenolide mycotoxin and of *Fusarium tricinctum* cultures in cattle. J Am Vet Med Assoc 160:1522–1526

Uhlig S, Torp M, Jarp J, Parich A, Gutleb A, Krska R (2004) Moniliformin in Norwegian grain. Food Addit Contam 21:598–606. https://doi.org/10.1080/02652030410001704258

Ukwuru MU, Ohaegbu CG, Muritala A (2017) An overview of mycotoxin contamination of foods and feeds. J Biochem Microb Toxicol 1(1):101

Umeda M, Tsutsui T, Saito M (1977) Mutagenicity and inducibility of DNA single-strand breaks and chromosome aberrations by various mycotoxins. Gann Jpn J Cancer Res 68(5):619–625

Umetsu N, Kaji J, Tamari K (1972) Investigation on the toxin production by several blast fungus strains and isolation of tenuazonic acid as a novel toxin. Agric Biol Chem 36:859–866

Umetsu N, Kaji J, Aoyama K, Tamari K (1974) Toxins in blast-diseased rice plants. Agric Biol Chem 38(10):1867–1874

Usleber E, Dade M, Schneider E, Dietrich R, Bauer J, Märtlbauer E (2008) Enzyme immunoassay for mycophenolic acid in milk and cheese. J Agric Food Chem 56(16):6857–6862. https://doi.org/10.1021/jf801063w

Van Asselt ED, Azambuja W, Moretti A, Kastelein P, De Rijk TC, Stratakou I, Van der Fels-Klerx HJ (2012) A Dutch field survey on fungal infection and mycotoxin concentrations in maize. Food Addit Contam A 29:1556–1565

Velasco M, Gilbert CA, Rutledge CO, McNay JL (1975) Antihypertensive effect of a dopamine beta hydroxylase inhibitor, bupicomide: a comparison with hydralazine. Clin Pharmacol Therapeut 18:145–153

Versilovskis A, De Saeger S (2010) Sterigmatocystin: occurrence in foodstuffs and analytical methods--an overview. Mol Nutr Food Res 54(1):136–147. https://doi.org/10.1002/mnfr.200900345

Vesonder RF, Gasdorf H, Peterson RE (1993) Comparison of the cytotoxicities of *Fusarium* metabolites and *Alternaria* metabolite AAL-toxin to cultured mammalian cell lines. Arch Environ Contam Toxicol 24:473–477

Viegas C, Nurme J, Piecková E, Viegas S (2020) Sterigmatocystin in foodstuffs and feed: aspects to consider. Mycology 11(2):91–104. https://doi.org/10.1080/21501203.2018.1492980

Wang C, Jiang Z, Yao J, Wu X, Sun L, Liu C, Duan W, Yan M, Sun L, Liu J, Zhang L (2008a) Participation of cathepsin B in emodin-induced apoptosis in HK-2 cells. Toxicol Lett 181:196–204

Wang YM, Wang HJ, Peng SQ (2008b) Lipid peroxidation and antioxidant defense impairment in the hearts of chick embryos induced by *in ovo* exposure to *Fusarium* mycotoxin butenolide. Toxicon 52(7):781–786. https://doi.org/10.1016/j.toxicon.2008.08.006

Wang HJ, Wang YM, Peng SQ (2009a) Repeated administration of a *Fusarium* mycotoxin butenolide to rats induces hepatic lipid peroxidation and antioxidant defense impairment. Food Chem Toxicol 47:633–637

Wang YM, Liu JB, Peng SQ (2009b) Effects of *Fusarium* mycotoxin butenolide on myocardial mitochondria *in vitro*. Toxicol Mech Methods 19(2):79–85. https://doi.org/10.1080/15376510802322802

Wang YM, Wang HJ, Peng SQ (2009c) *In ovo* exposure of a *Fusarium* mycotoxin butenolide induces hepatic and renal oxidative damage in chick embryos, and antioxidants provide protections. Toxicol in Vitro 23(7):1354–1359. https://doi.org/10.1016/j.tiv.2009.06.028

Wang J, Huang S, Xing L, Cui J, Tian Z, Shen H, Jiang X, Yan X, Wang J, Zhang X (2015) Sterigmatocystin induces G1 arrest in primary human esophageal epithelial cells but induces G2 arrest in immortalized cells: key mechanistic differences in these two models. Arch Toxicol 89(11):2015–2025

Wang M, Jiang N, Xian H, Wei D, Shi L, Feng X (2016) A single-step solid phase extraction for the simultaneous determination of 8 mycotoxins in fruits by ultra-high performance liquid chromatography tandem mass spectrometry. J Chromatogr A 1429:22–29

Wareing PW, Westby A, Gibbs JA, Allotey LT, Halm M (2001) Consumer preferences and fungal and mycotoxin contamination of dried cassava products from Ghana. Int J Food Sci Technol 36 (1):1–10

van Warmelo KT, Marasas WFO (1972) *Phomopsis leptostromiformis*: the causal fungus of lupinosis, a mycotoxicosis, in sheep. Mycologia 64(2):316–324. https://doi.org/10.1080/00275514.1972.12019266

Warth B, Parich A, Atehnkeng J, Bandyopadhyay R, Schuhmacher R, Sulyok M, Krska R (2012) Quantitation of mycotoxins in food and feed from Burkina Faso and Mozambique using a modern LC-MS/MS multitoxin method. J Agric Food Chem 60(36):9352–9363

Weber J, Vaclavikova M, Wiesenberger G, Haider M, Hametner C, Fröhlich J, Berthiller F, Adam G, Mikula H, Fruhmann P (2018) Chemical synthesis of culmorin metabolites and their biologic role in culmorin and acetyl-culmorin treated wheat cells. Org Biomol Chem 16(12):2043–2048

Wehner FC, Thiel PG, Rand MD (1979) Mutagenicity of the mycotoxin emodin in the *Salmonella*/microsome system. Appl Environ Microbiol 37(3):658–660

Weidenbörner M (2001) Encyclopedia of food mycotoxins. Springer Science & Business Media, New York, NY

Wells JM, Cole RJ, Kirksey JW (1975) Emodin, a toxic metabolite of *Aspergillus wentii* isolated from weevil-damaged chestnuts. Appl Microbiol 30(1):26–28. https://doi.org/10.1128/am.30.1.26-28.1975

Wilson DM (1976) Patulin and penicillic acid. In: Rodrick JV (ed) Mycotoxins and other fungal related problems, Advances in chemistry series, vol 149. American Chemical Society, Washington, DC, p 90

Woelflingseder L, Adam G, Marko D (2020) Suppression of trichothecene-mediated immune response by the *Fusarium* secondary metabolite butenolide in human colon epithelial cells. Front Nutr 7:127. https://doi.org/10.3389/fnut.2020.00127

Woody MA, Chu FS (1992) Toxicology of *Alternaria* mycotoxins. In: Chełkowski J, Visconti A (eds) Topics in secondary metabolism. Volume 3. *Alternaria*: biology, plant diseases and metabolites. Elsevier, New York, NY, pp 409–434

Yang HY, Wang YM, Peng SQ (2010) Metallothionein-I/II null cardiomyocytes are sensitive to *Fusarium* mycotoxin butenolide induced cytotoxicity and oxidative DNA damage. Toxicon 55: 1291–1296

Yassin MA, Ama E-S, Moslem MA, El-Naggar MA (2013) Mycobiota of almond seeds and the toxigenicity of some involved genera. Life Sci J 10(4):1088–1093

Yin ES, Rakhmankulova M, Kucera K, de Sena Filho JG, Portero CE, Narváez-Trujillo A, Holley SA, Strobel SA (2015) Fusaric acid induces a notochord malformation in zebrafish via copper chelation. BioMetals 28:783–789

Yoshinari T, Takeuchi H, Kosugi M, Taniguchi M, Waki M, Hashiguchi S, Fujiyoshi T, Shichinohe Y, Nakajima M, Ohnishi T, Hara-Kudo Y, Sugita-Koshini Y (2019) Determination of sterigmatocystin in foods in Japan: method validation and occurrence data. Food Addit Contam A Chem Anal Cont Expos Risk Assess 36(9):1404–1410. https://doi.org/10.1080/19440049.2019.1628359

Yu J, Jurick WM, Cao H, Yin Y, Gaskins VL, Losada L, Zafar N, Kim M, Bennett JW, Nierman WC (2014) Draft genome sequence of Penicillium expansum strain R19, which causes postharvest decay of apple fruit. Genome Announc 2(3):e00635–e00614

Zheng YF, Liu CF, Lai WF, Xiang Q, Li ZF, Wang H, Lin N (2014) The laxative effect of emodin is attributable to increased aquaporin 3 expression in the colon of mice and HT-29 cells. Fitoterapia 96:25–32. https://doi.org/10.1016/j.fitote.2014.04.002

Zinedine A, Meca G, Mañes J, Font G (2011) Further data on the occurrence of *Fusarium* emerging mycotoxins enniatins (A, A1, B, B1), fusaproliferin and beauvericin in raw cereals commercialized in Morocco. Food Control 22:1–5. https://doi.org/10.1016/j.foodcont.2010.05.002

Zinedine A, Fernández-Franzón M, Mañes J, Manyes L (2017) Multi-mycotoxin contamination of couscous semolina commercialized in Morocco. Food Chem 214:440–446. https://doi.org/10.1016/j.foodchem.2016.07.098

Zouaoui N, Mallebrera B, Berrada H, Abid-Essefi S, Bacha H, Ruiz MJ (2016) Cytotoxic effects induced by patulin, sterigmatocystin and beauvericin on CHO-K1 cells. Food Chem Toxicol 89: 92–103

Chapter 5
Mushroom Toxins and Health Hazards Associated with Them

5.1 Introduction

Mushrooms are fungi having spore-bearing and fleshy fruiting bodies. They belong to Ascomycota or Basidiomycota (Miles and Chang 2004; Chang 2008). Since ancient times, mushrooms have been consumed as food components, especially in some ethnic populations (Barman et al. 2017). They are rich in protein, carbohydrates, and fibres. Because of their high nutritional value and subtle taste, mushrooms are becoming an important part of the human diet. Several species of mushrooms are also known for their medicinal value. Their secondary metabolites or bioactive compounds have been used as therapeutics. However, few of them are poisonous and are mycotoxin producers. About 120 mushroom toxins have been reported (Yin et al. 2019). Mushroom poisoning usually occurs due to the consumption of poisonous mushrooms having striking similarities with edible mushrooms. Moreover, the accidental consumption of wild mushrooms or foraged mushrooms can lead to hazardous health effects.

5.2 Classification of Toxic Mushrooms

More than 100 species of poisonous mushrooms have been reported to date (Patocka et al. 2021). The syndromic classification of toxic mushrooms was published in 1994. Since then, many new syndromes have been added to mushroom poisoning. Based on specific clinical symptoms, White et al. (2019) classified them into six groups:

Group 1: Cytotoxic mushroom toxins

They cause organ toxicity and are subdivided into three groups:

K. Singh, A. Kumari, *Mycotoxins and Mycotoxicoses*,
https://doi.org/10.1007/978-981-19-2370-8_5

(a) Primary hepatotoxic (e.g. amatoxins).
(b) Primary nephrotoxic.

- Early (amino hexadienoic acid, AHDA).
- Delayed (orellanines).

Group 2: Neurotoxic mushroom (toxins primarily neurotoxic)

This category includes the following:

(a) Hallucinogenic (psilocybins and related toxins).
(b) Autonomic toxicity (muscarines).
(c) Central nervous system-toxicity (muscimol/ibotenic acid).
(d) Morel neurologic syndrome (toxins from *Morchella* spp.)

Group 3: Myotoxic mushroom causing poisoning syndrome (primary symptom rhabdomyolysis); this includes two subgroups:

(a) Rapid response (*Russula* spp.)
(b) Delayed response (*Tricholoma* spp.)

Group 4: Affected endocrine and/or metabolism-related process

(a) Gamma-aminobutyric acid (GABA)-blocking mushrooms (gyromitrins).
(b) Coprines (disulfiran-like).
(c) Polyporic mushrooms (polyporic acid).
(d) Trichothecene mushrooms (*Podostroma* spp.)
(e) Hypoglycemic mushroom (*Trogia venenata*).
(f) Hyperprocalcitoninemia mushrooms (*Boletus satanas*).
(g) Pancytopenic mushrooms (*Ganoderma neojaponicum*).

Group 5: GI-irritant mushrooms.
Group 6: Adverse sundry reactions (which are not included in the above five groups)

(a) Dermatitis caused by Shiitake mushroom (*Lentinula edodes*).
(b) Erythromelagic mushrooms (*Clitocybe acrometagia*).
(c) Paxillus syndrome caused by *Paxillus involutus.*
(d) Encephalopathy syndrome caused by *Pleurocybella porrigens.*

Although most of the mushroom toxins have been characterized, some toxins remain unidentified.

The major mushroom toxins are described in the following sections.

Fig. 5.1 *Gyromitra gigas* (a giants false morel)—five specimens. (This figure is reprinted with permission from Alan H Hall, Colorado School of Public Health, Pediatrics in Review, vol. 8 no. 10 April 1987)

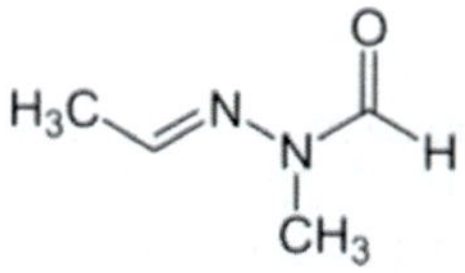

Fig. 5.2 Chemical structure of gyromitrin. (Adapted from By Jü, Own work, Public Domain, https://commons.wikimedia.org/w/index.php?curid=11065616)

5.3 Gyromitrin

Gyromitrin is produced by false morel *Gyromitra esculenta* and giants false morel, *Gyromitra gigas* (Fig. 5.1). It was isolated in 1967 as acetaldehyde *N*-methyl-*N*-formylhydrazone (List and Luft 1967).

The heat treatment and other physiological conditions decompose gyromitrin into a more toxic compound MFH (*N*-methyl-*N*-formlhydrazin), which also decomposes to monomethylhydrazin (MH) (Pyysalo and Niskanen 1977). The molecular formula of gyromitrin is $C_4H_8N_{2O}$. Its molecular weight is 100.12. Its IUPAC name is *N*-[(E)-ethylideneamino]-*N*-methylformamide (Fig. 5.2).

5.3.1 Toxicity

LD_{50} of gyromitrin is 20–50 mg/kg. The clinical symptoms of gyromitrin poisoning include neurological seizures, convulsions, GI tract and hepatic disorders, alteration in kidney functioning, and death (Yin et al. 2019). Gyromitrin is heat-labile; thus, consumption of raw and incompletely cooked false morel mushrooms can be fatal. MFH causes hepatotoxicity, while MH has been found carcinogenic (Braun et al. 1981).

5.4 Amanitin

Amatoxins and phallotoxins are cyclic octapeptides and heptapeptides. They are produced by *A. phalloides*, *A. virosa* (Fig. 5.3), and *A. verna*.

The octapeptides are the most lethal toxins. The α-amanitin and phallacidin are biosynthesized by NRPSs (non-ribosomal peptide synthetases) and encoded by AMA1, PHA1, and MSDIN gene families (Luo et al. 2012). The cyclopeptides are of the following types:

(a) Slow and violent-acting amatoxins which are highly toxic (LD_{50} 0.4–0.8 mg/kg in albino mice). They cause hepatocellular and nephrotoxicity, thus causing death within 2–4 days of consumption (Karlson-Stiber and Persson 2003). The molecular formula of α-amanitin is $C_{39}H_{54}N_{10}O_{14}S$ and molecular weight is 919.0 (Pubchem) whereas IUPAC name is 2-[33-butan-2-yl-12-(3,4-dihydroxybutan-2-yl)-8,21-dihydroxy-2,5,10,13,26,29,32,35,38-nonaoxo-

Fig. 5.3 *Amanita virosa* (**a**) death cup or volva at the base of stem without ring; (**b**) volva with ring of death or annulus. (This figure is reprinted with permission from Alan H Hall, Colorado School of Public Health, Pediatrics in Review, vol. 8 no. 10 April 1987)

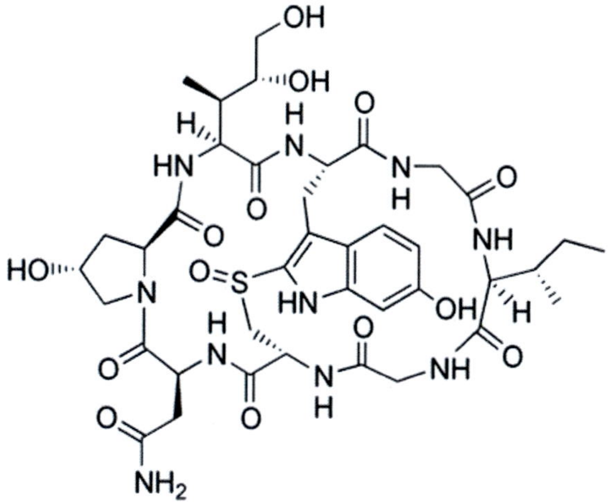

Fig. 5.4 Chemical structure of α-amanitin. (Adapted from Edgar181, Own work, Public Domain, https://commons.wikimedia.org/w/index.php?curid=5909292)

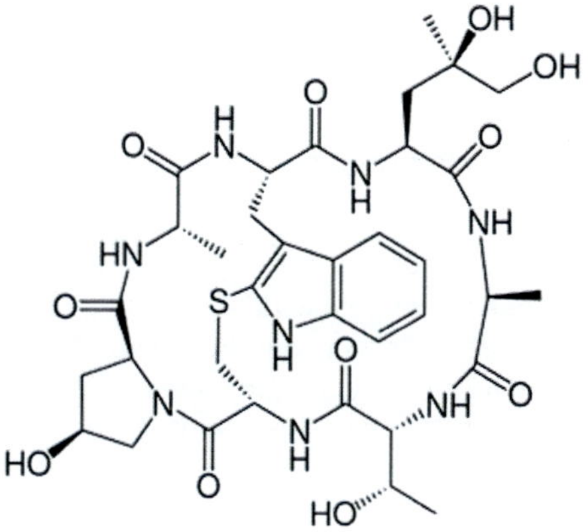

Fig. 5.5 Chemical structure of phalloidin. (Adapted from Chem Sim 2001, Own work, Public Domain, https://commons.wikimedia.org/w/index.php?curid=94089573)

26λ4-thia-3,6,11,14,24,28,31,34,37 nonazapentacyclo [13.12.11.16,9.017,25.018,23]nonatriaconta-17(25),18(23),19,21-tetraen-4-yl] acetamide (Fig. 5.4).

(b) Quick-acting phallotoxins (Fig. 5.5) are less toxic than amatoxins (LD_{50} 1–20 mg/kg in albino mice). The death occurs within 2–5 h. They disrupt the functioning of the cytoskeleton (Garcia et al. 2015).

(c) Virotoxins are less toxic (Vetter 1998). Their symptomatology in mice includes haemorrhagic hepatic necrosis (Loranger et al. 1985).

(d) Antamanides are non-toxic to humans and animals and act as an antidote to phallotoxins (Wieland et al. 1968).

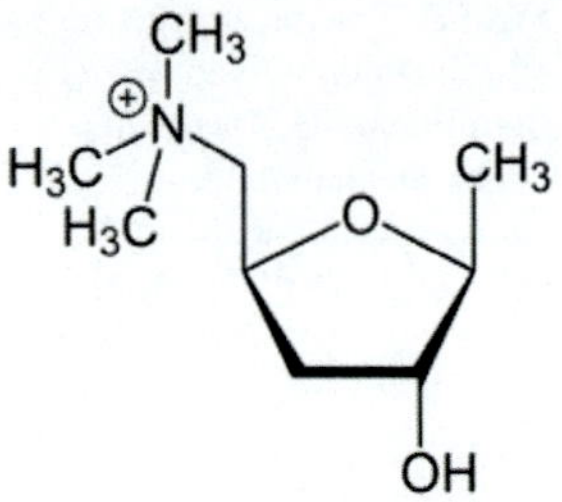

Fig. 5.6 Chemical structure of muscarine. (By Yikrazuul, Own work, Public Domain, https://commons.wikimedia.org/w/index.php?curid=7444182)

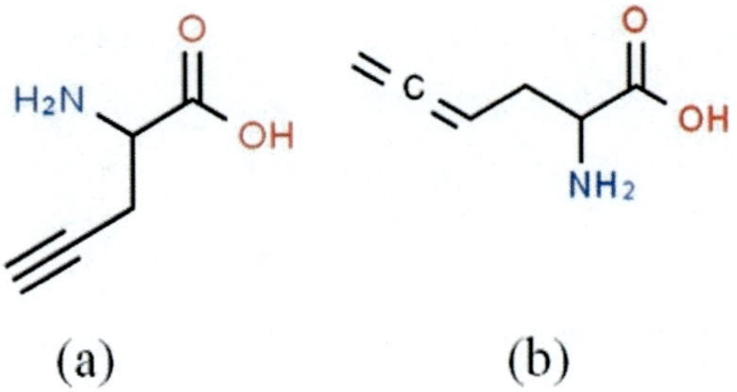

Fig. 5.7 Chemical structure of (**a**) L-2-amino-4-pentynoic acid (http://www.chemspider.com/Chemical-Structure.86268.html) and (**b**) L-2-amino-4,5-hexadienoic acid. (http://www.chemspider.com/Chemical-Structure.134192.html)

5.5 Isoxazoles

The isoxazoles are the products of *Amanita muscaria* and *A. pantherina*. Three isoxazoles viz. ibotenic acid (IBO), muscazone, and muscimol (MUS) have been reported. The symptoms of isoxazoles poisoning are mycoatropinic. MUS (Fig. 5.6) is a hallucinogenic compound, while muscazone induces mild pharmacological activities only (Voynova et al. 2020).

5.5.1 *Toxicity*

The symptoms of poisoning such as dizziness, space distortion, visual and auditory anaesthesia, and severe damage to neurons appear half an hour to 2 h after consumption. IBO with MUS causes brain lesions and neuron damage (Schwarcz et al. 1979). The euphoria with anxiety, inebriation, and violent behaviour is common in isoxazoles poisoning (Piqueras 2000).

5.6 Amino Acids

In 1978, the poisoning of two amino acids from *A. abrupta* was reported in Nagano, Japan. Later, Yamaura et al. (1986) identified them as L-2 amino-4-pentynoic acid and L-2-amino-4,5-hexadienoic acid. The molecular formula of L-2 amino-4-pentynoic acid is C_5H_7NO (Fig. 5.7a). The chemical formula of L-2-amino-4,5-hexadienoic (Fig 5.7b) acid is $C_6H_9NO_2$. Its molecular weight is 127.14.

5.6.1 *Toxicity*

The symptoms of intoxication include vomiting, dehydration, and diarrhoea. They were reported to cause hepatocellular necrosis in mice (Yamaura et al. 1986).

5.7 Orellanine or Orellanine–Trifluoroacetic Acid Complex

Orellanine and its derivatives are produced by mushrooms *Cortinarius orellanus* and *Clerodendrum speciosissimum* that exhibit striking similarities with edible mushrooms *Cantharellus tubaeformis* and *Cantharellus cibarius*. The molecular formula of orellanine is $C_{10}H_8N_2O_6$ and its IUPAC name is 3,3′,4,4′-Tetrahydroxy-2,2-′-bipyridine-*N*,*N*′-dioxide (Fig. 5.8). The molecular weight of orellanine is 152.18 (https://pubchem.ncbi.nlm.nih.gov/compound/89579).

5.7.1 *Toxicity*

The toxicity of orellanine was reported in 1957. The accidental consumption of *C. orellanus* and *C. speciosissimus* in humans causes renal toxicity and pathological changes in the spleen and liver. The toxicity of orellanine is reported in animals like guinea pigs, cats, and mice (Yin et al. 2019). The latent period of orellanine poisoning is 36 h (which induces delayed response). The symptomatology includes nausea, vomiting, increased thirst, chills, lower back pain, and oliguria, occasionally followed by dieresis (Hall et al. 1987). Orellanine derivatives can induce oxidative

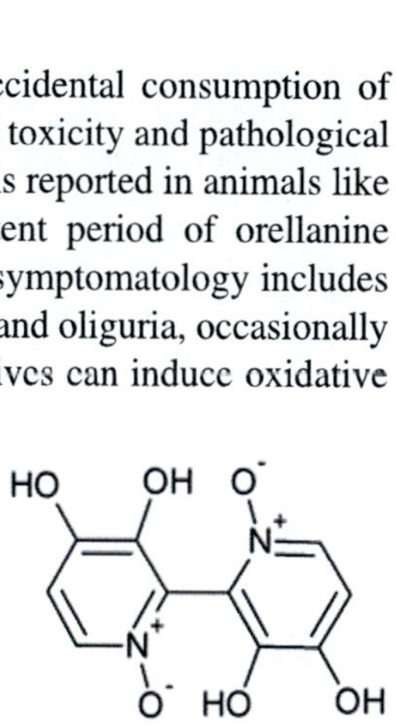

Fig. 5.8 Chemical structure of orellanine. (Adapted from LHcheM, Own work, CC BY-SA 3.0, https://commons.wikimedia.org/w/index.php?curid=44617750)

stress causing damage to DNA, RNA, and proteins (Cohen-Addad et al. 1987). There is no antidote for orellanine poisoning. However, charcoal haemoperfusion is recommended to combat the toxin (Hall et al. 1987).

5.8 Hydroxyindole Derivatives

The indole derivatives psilocybin, baeocystin, norbaeocystin, and norpsilocin are produced by psychedelic mushrooms *Psilocybe mexicana*, *P. baeocystin*, *P. cubensis*, *P. semilanceata*, *P. bohemica*, and *P. baeocistis*. These compounds have been in use by some Native American groups for a long time (Leung and Paul 1968; Bogusz et al. 1998; Sherwood et al. 2020). Aeruginascin is a naturally occurring toxin of *Inocybe aeruginascens* and *Pholiotina cyanopus* (Sherwood et al. 2020).

5.8.1 Psilocybin

The molecular formula of psilocybin is $C_{12}H_{17}N_2O_4P$ and its molecular weight is 284.25 (Fig. 5.9). It activates the serotonin receptors causing hallucinations (https://www.medicalnewstoday.com/articles/308850). Its poisoning in children has been reported to cause fever with convulsions. In adults, the intoxication symptoms start within 30 min to 2 h after consumption of psilocybin. It induces vomiting, anxiety, delusions, visual distortion, confusion, and acute schizophrenia (https://www.livsmedelsverket.se/en/food-and-content/food/mushrooms-and-mushroom-toxins?AspxAutoDetectCookieSupport=1).

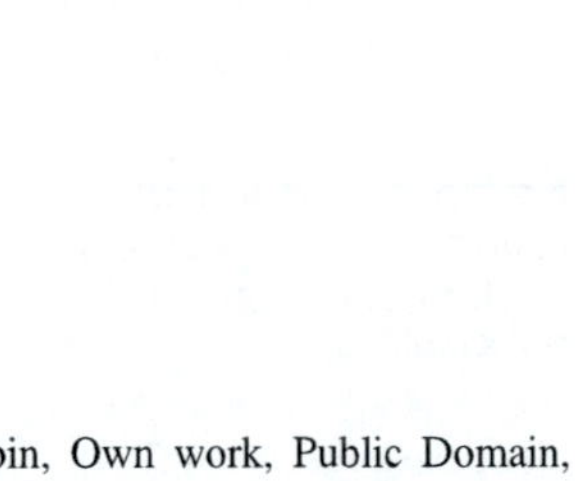

Fig. 5.9 Chemical structure of psilocybin. (Adapted from en:User:Cburnett, Own work in kdrawchem and tweaked with Inkscape, CC BY-SA 3.0, https://commons.wikimedia.org/w/index.php?curid=5785906)

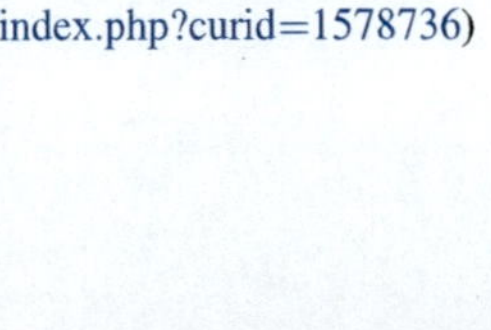

Fig. 5.10 Chemical structure of psilocin. (Adapted from Harbin, Own work, Public Domain, https://commons.wikimedia.org/w/index.php?curid=1578736)

5.8.2 Psilocin

Psilocin ($C_{12}H_{16}N_2O$) is an indole hallucinogen and a metabolite of psilocybin (Fig. 5.10). Its molecular weight is 204.27. The IUPAC name of psilocin is 3-[2-(dimethylamino)ethyl]-1H-indol-4-ol (Pubchem).

It can alter neurotransmission, thereby inducing psychoactive effects. It has a high affinity with the 5-HT_{2A} serotonin binding receptor (Catlow et al. 2013) and can inhibit pyramidal cell activity (Puig et al. 2005). In dogs, psilocin induces vocalization, ataxia, aggression, hyperthermia, and nystagmus (Kirwan 1990). Its effects on humans include incoordination, increased heart rate, visual hallucinations, tremors, increased blood pressure, intensified hearing, and pupil dilation (Puschner 2018).

5.8.3 Baeocystin

Baeocystin is an analogue of psilocybin (Leung and Paul 1968). Its molecular formula is $C_{11}H_{15}N_2O_4P$, while its molecular weight is 270.22 (Fig. 5.11). Baeocystin is *N*-methyltryptamine (a tryptamine alkaloid) and its IUPAC name is [3-[2-(methylamino)ethyl]-1H-indol-4-yl] dihydrogen phosphate (Pubchem). Baeocystin is as psychoactive as psilocybin (Gartz 1997). Its psychedelic activity was not observed in mice (Sherwood et al. 2020).

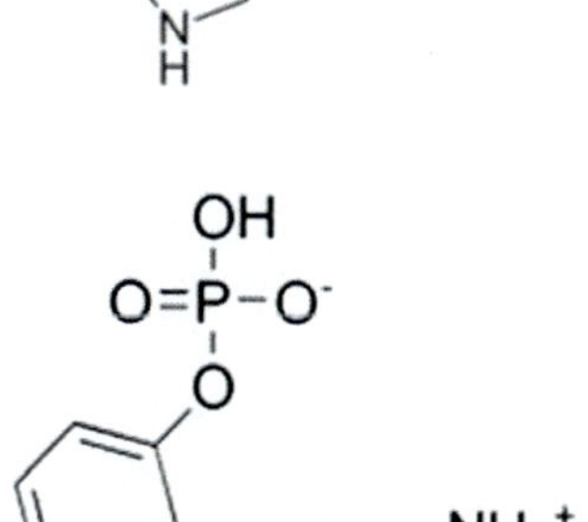

Fig. 5.11 Chemical structure of baeocystin. (Adapted from Jü, Own work, CC0, https://commons.wikimedia.org/w/index.php?curid=20375448)

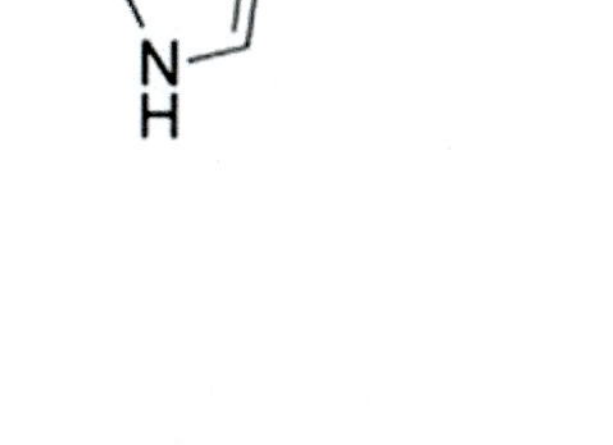

Fig. 5.12 Chemical structure of norbaeocystin. (Adapted from User Cacycle on en.wikipedia, Public Domain, https://commons.wikimedia.org/w/index.php?curid=1258159)

5.8.4 Norbaeocystin

Norbaeocystin is a hallucinogen and a tryptamine alkaloid (Pubchem). Like baeocystin, it is also an analogue of psilocybin (Sherwood et al. 2020). The molecular formula of norbaeocystin is $C_{10}H_{13}N_2O_4P$ and its molecular weight is 256.19. The IUPAC name of norbaeocystin is [3-(2-aminoethyl)-1H-indol-4-yl] dihydrogen phosphate (Fig. 5.12) (Pubchem). It has poor BBB permeability and exerts psychedelic effects (Pajouhesh and Lenz 2005).

5.8.5 Aeruginascin

Aeruginascin is a minor component produced by a poisonous mushroom *Inocybe aeruginascens*. Gartz isolated it in 1987 (Gartz 1987). Aeruginascin is the *N*-trimethyl analogue of psilocybin (Chadeayne et al. 2020). The molecular formula of aeruginascin is $C_{13}H_{19}N_2O_4P$ (Fig. 5.13). Its molecular weight is 298.27. The IUPAC name of aeruginascin is [3-[2-(trimethylazaniumyl)ethyl]-1H-indol-4-yl] hydrogen phosphate.

Jensen et al. (2006) described its closed relatedness with bufotenidine (5-HTQ) (a toxin from frog skin). However, its strong binding was observed at the 5-HT_2 receptors, not at the 5-HT_3 receptors (Chadeayne et al. 2020). Aeruginascin perhaps alters the psychedelic effects to a euphoric feeling (Gartz 2008). The significant role of aeruginascin and its metabolite, 4-hydroxy-*N*,*N*,*N*-trimethyltryptamonium (4-OH-TMT) in WLP (wood lover paralysis) was reported by Sherwood et al. (2020).

5.9 Acromelic Acids (ACROs)

ACROs are produced by *Clitocybe* spp. (Fig. 5.14) like *C. acromelaga*.

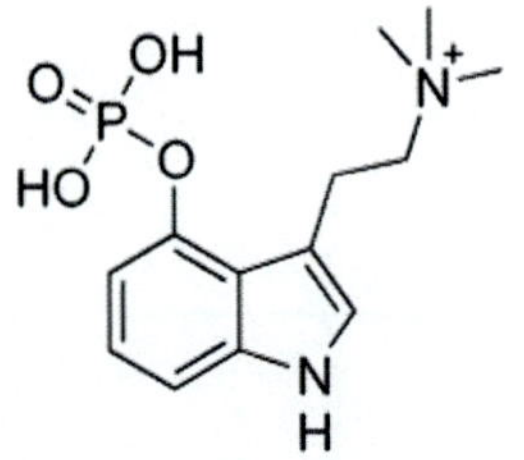

Fig. 5.13 Chemical structure of aeruginascin. (Adapted from Meodipt at English Wikipedia, Transferred from en.wikipedia to Commons by Edgar181 using CommonsHelper., Public Domain, https://commons.wikimedia.org/w/index.php?curid=19034969)

Fig. 5.14 *Clitocybe* sp. (This figure is reprinted with permission from Alan H Hall, Colorado School of Public Health, Pediatrics in Review, vol. 8 no. 10 April 1987)

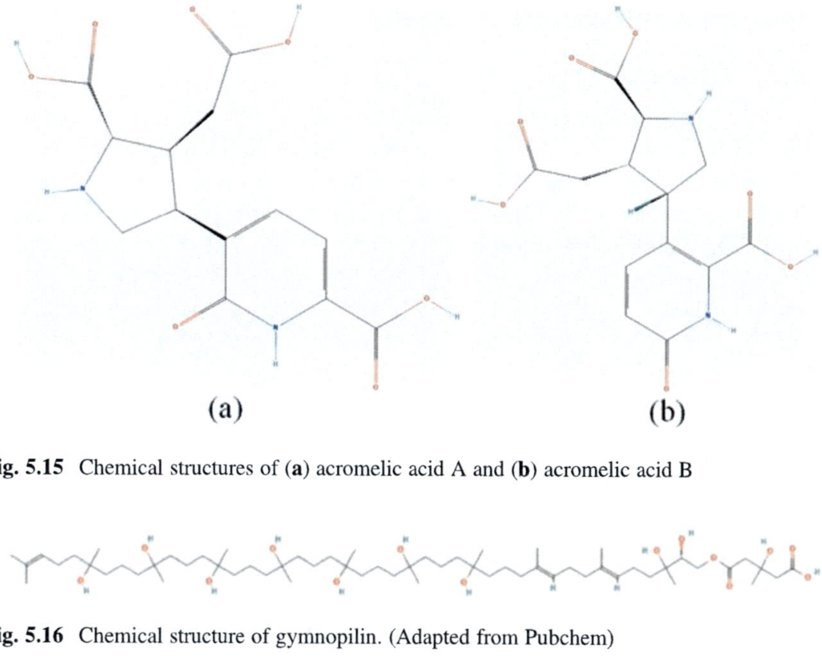

Fig. 5.15 Chemical structures of (**a**) acromelic acid A and (**b**) acromelic acid B

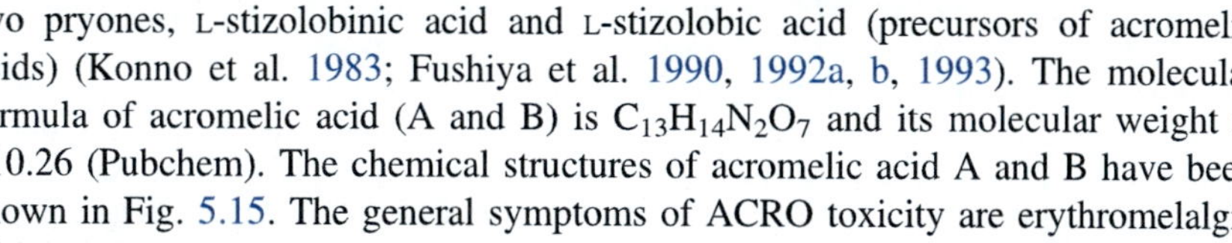

Fig. 5.16 Chemical structure of gymnopilin. (Adapted from Pubchem)

ACROs are kainoids. This group of toxins is composed of acromelic acid A–E, two amino acids, β-cyano-L-alanine and *N*-(γ-L-glutamyl)-β-cyano-L-alanine, and two pryones, L-stizolobinic acid and L-stizolobic acid (precursors of acromelic acids) (Konno et al. 1983; Fushiya et al. 1990, 1992a, b, 1993). The molecular formula of acromelic acid (A and B) is $C_{13}H_{14}N_2O_7$ and its molecular weight is 310.26 (Pubchem). The chemical structures of acromelic acid A and B have been shown in Fig. 5.15. The general symptoms of ACRO toxicity are erythromelalgia with burning pain, neurotoxicity, and neuroexcitatory activity. The administration of acromelic A induces clonic seizures and other abnormal behaviours in rats and allodynia in mice (Minami et al. 2004; Taguchi et al. 2009).

5.10 Gymnopilins

Gymnopilin is produced by the fruiting bodies of big laugher mushroom *Gymnopilus spectabilis* (Yin et al. 2019). The IUPAC name of gymnopilin is 3-hydroxy-3-methyl-5-[(2R,6E,10E)-2,3,15,19,23,27,31,35,39-nonahydroxy-3,7,11,15,19,23,27,31,35,39,43undecamethyltetratetraconta-6,10,42-trienoxy]-5-oxopentanoic acid. Its molecular formula and molecular weight are $C_{61}H_{114}O_{14}$ and 1071.5, respectively (Fig. 5.16). Nozoe et al. (1983) elucidated the structure of this mushroom toxin (Nozoe et al. 1983).

5.10.1 *Toxicity*

Gymnopilin causes excessive laughing. It can cross the blood–brain barrier and can stimulate the vasomotor centre. This mushroom toxin has a depolarizing effect on CNS (Tanaka et al. 1993). Gymnopilin acts on the spinal cord and can inhibit nicotine-like acetylcholine receptors (Kayano et al. 2014). It also induces

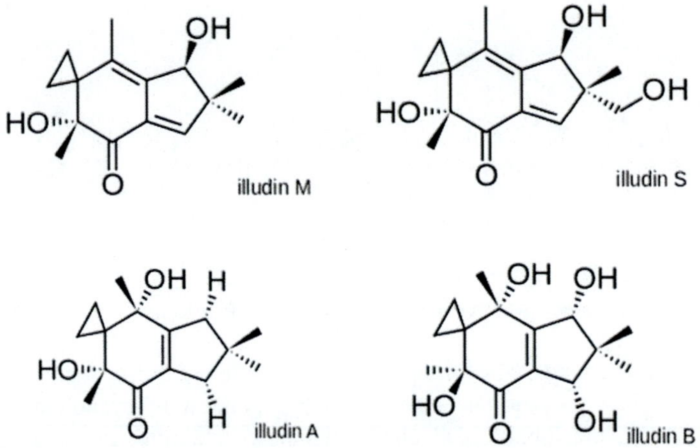

Fig. 5.17 Chemical structure of illudins. (Adapted from Fvasconcellos 18:57, 9 December 2007 (UTC), Own work, after Padwa A, Sandanayaka VP, Curtis EA (1994). Synthetic studies toward Illudins and Ptaquilosin. A highly convergent approach via the dipolar cycloaddition of carbonyl ylides;. J Am Chem Soc 116 (6): 2667–8. doi:10.1021/ja00085a076, Public Domain, https://commons.wikimedia.org/w/index.php?curid=3204459)

vasodilation in smooth muscles of the aorta, a temporary increase in blood pressure, and cardiac output in rats (Nishio et al. 2012).

5.11 Illudins

Illudin S and illudin M are sesquiterpenes produced by the fruiting bodies of jack-o'-lantern mushroom (*Omphalotus illudens*) and moonlight mushroom (*O. Japonicas*) (McMorris and Anchel 1965). The molecular formula of illudin S is $C_{15}H_{20}O_4$ and its molecular weight is 264.32 (Fig. 5.17). Its IUPAC name is (1R,2S,5R)-1,5-dihydroxy-2-(hydroxymethyl)-2,5,7-trimethylspiro[1H-indene-6,1-′-cyclopropane]-4-one (Pubchem). The symptoms of illudin intoxication are vomiting, indigestion, diarrhoea, cramping, and stomach ache (Yin et al. 2019).

5.12 Hebevinosides

Hebevinosides I to XIV are toxic compounds of poisonous mushroom *Hebeloma vinosophyllum* (Fujimoto et al. 1986), while toxic terpenoids HS-A, HS-B, and HS-C were isolated from poisonous mushrooms *He. spoliatum*, *He. sinapizans*, and *He. crustuliniforme* (Fujimoto et al. 1992). The molecular weight of

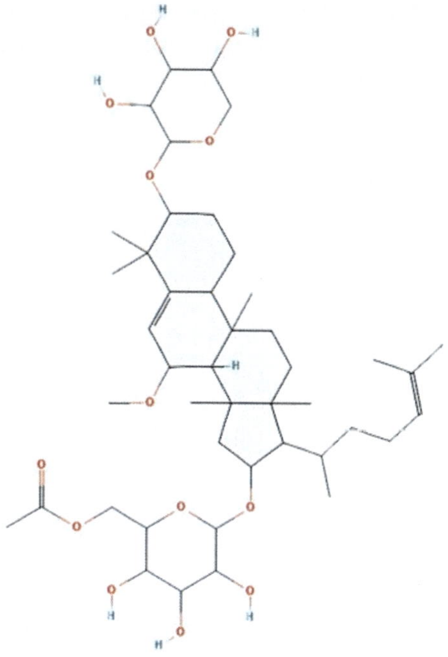

Fig. 5.18 Chemical structure of hebevinoside I. (Adapted from Pubchem)

hebevinoside I is 809.0. Its molecular formula is $C_{44}H_{72}O_{13}$ (Fig. 5.18). The IUPAC name of this toxic compound is [3,4,5-trihydroxy-6-[[7-methoxy-4,4,9,13,14-pentamethyl-17-(6-methylhept-5-en-2-yl)-3-(3,4,5-trihydroxyoxan-2-yl)oxy-2,3,7,8,10,11,12,15,16,17-decahydro-1H-cyclopenta[a]phenanthren-16-yl]oxy]oxan-2-yl]methyl acetate (Pubchem). Hebevinosides I to XIV are neurotropic and cause gastrointestinal toxicity and neurotoxicity in mice. HS-A, HS-B, and HS-C are cytotoxic (De Bernardi et al. 1983).

5.13 Cycloprop-2-ene Carboxylic Acid

Matsuura et al. (2009) isolated cycloprop-2-ene carboxylic acid (Fig. 5.19) responsible for the poisoning of mushroom *Russula subnigricans*. Cycloprop-2-ene carboxylic acid ($C_4H_4O_2$) is an unstable compound. Its molecular weight is 84.07, and its IUPAC name is cycloprop-2-ene-1-carboxylic acid (Pubchem).

5.13.1 Toxicity

R. subnigricans poisoning was reported for the first time in Japan in 1954 (Yin et al. 2019). Cycloprop-2-ene carboxylic acid is a GI toxin. Its LD_{100} value in mice is 2.5 mg/kg BW. Its ingestion causes rhabdomyolysis with acute renal injury, pulmonary oedema, respiratory failure, severe electrolyte disturbance, cardiogenic shock, and ventricular tachycardia (Cho and Han 2016). The early management of cycloprop-2-ene carboxylic acid poisoning includes fast and voluminous repletion of fluids and removal of myoglobin from the blood by hemodiafiltration (Cho and Han 2016).

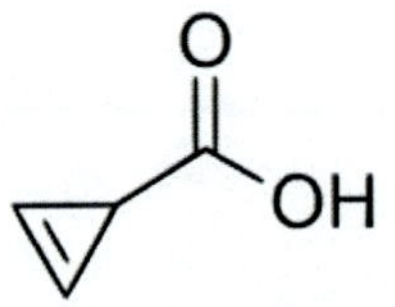

Fig. 5.19 Chemical structure of cycloprop-2-ene carboxylic acid. (Adapted from Mfernflower, Own work, Public Domain, https://commons.wikimedia.org/w/index.php?curid=42906099)

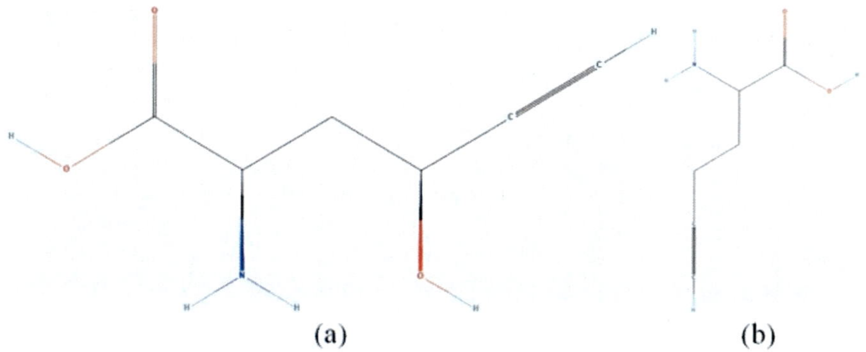

Fig. 5.20 Chemical structure of (a) 2R-amino-4S-hydroxy-5-hexynoic acid and (b) 2R-amino-5-hexynoic acid. (Source: Pubchem)

5.14 2R-Amino-4S-Hydroxy-5-Hexynoic Acid and 2R-Amino-5-Hexynoic Acid

The mushroom *Trogia venenata* has been associated with sudden unexplained death cases in Yunnan Province, China (Shi et al. 2012). Later, two toxins, 2R-amino-4S-hydroxy-5-hexynoic acid and 2R-amino-5-hexynoic acid (Fig. 5.20a, b), were found to be responsible for the toxicity of *Trogia venenata* (Zhou et al. 2012). The molecular weights of 2R-amino-4S-hydroxy-5-hexynoic acid ($C_6H_9NO_3$) and 2R-amino-5-hexynoic acid ($C_6H_9NO_2$) are 143.14 and 127.14, respectively. Their IUPAC names are (2R, 4S)-2-amino-4-hydroxyhex-5-ynoic acid and 2-aminohex-5-ynoic acid, respectively. The toxicological profile of these two toxic amino acids includes rhabdomyolysis and sudden death due to cardiac problems. The extract of *T. venenata* was reported to induce hypoglycaemia and death in mice (Shi et al. 2012).

5.15 Saponaceolides

Tricholoma spp. have been associated with delayed rhabdomyolysis cases. The two saponaceolides B and M were isolated from fruiting bodies of *Tricholoma terreum* and *Tr. equestre* (Bedry et al. 2001). The molecular weight of saponaceolide B ($C_{30}H_{46}O_6$) is 502.7. Its IUPAC name is (3E)-3-[2-[(1S,3R)-3-[2-[(1R,4R)-1-hydroxy-4,6,6-trimethylspiro[2,5-dioxabicyclo[2.2.2]octane-3,6′-oxane]-3′-yl] ethyl]-2,2-dimethyl-6-methylidenecyclohexyl]ethylidene]oxolan-2-one (Pubchem). The molecular weight of saponaceolide M is 574.7. Its molecular formula is $C_{32}H_{46}O_9$. The IUPAC name of saponaceolide M is [(2E)-1-[(1S,3R)-3-[2-[(1R,3S,3′S,4S)-1-hydroxy-4,6,6-trimethyl-7-oxospiro[2,5-dioxabicyclo[2.2.2]

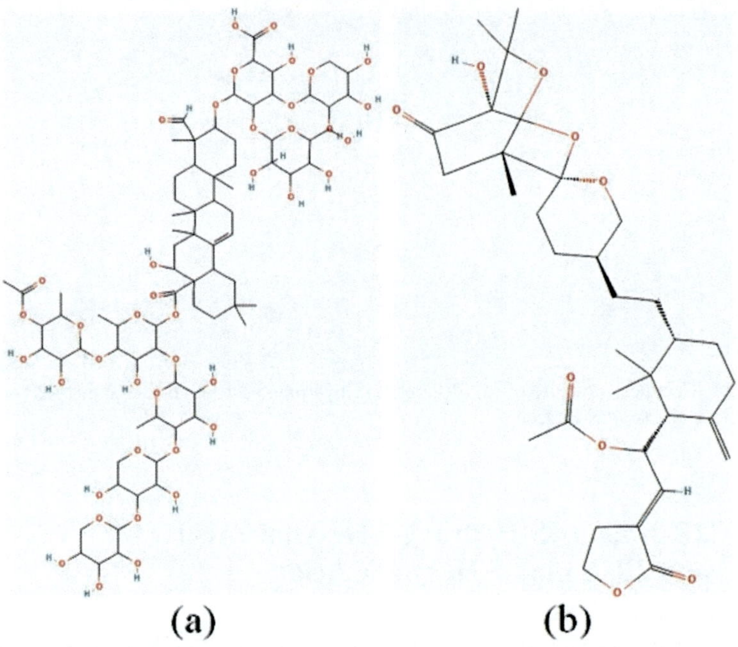

Fig. 5.21 Chemical structure of (**a**) saponaceolide B and (**b**) saponaceolide M

octane-3,6′-oxane]-3′-yl]ethyl]-2,2-dimethyl-6-methylidenecyclohexyl]-2-(2-oxooxolan-3-ylidene)ethyl] acetate (Fig. 5.21a, b) (Pubchem). Apart from these, saponaceolides E, F, and G and saponaceoic acids I, II and III were isolated from *T. saponaceum* (Yoshikawa et al. 2002). Saponaceolides-induced elevation in serum CK level caused toxicity in mice (Davoli et al. 2016).

5.16 Rare Mushroom Toxins

5.16.1 Trichothecenes

Podostroma cornu-damae is a rare fungus found in Java and Japan. It is commonly known as poison fire coral. This deadly fungus produces three toxic trichothecenes: verrucarin J, roridin E, and satratoxin H and their derivatives. The molecular weights of verrucarin J ($C_{27}H_{32}O_8$), roridin E ($C_{29}H_{38}O_8$), and satratoxin H ($C_{29}H_{36}O_9$) are 484.5, 514.6, and 528.6, respectively (Fig. 5.22a–c) (Pubchem). The accidental ingestion of *P. cornu-damae* can cause bicytopenia, hair loss, voluntary movement-related problems, and GI tract disorders (Kim et al. 2016).

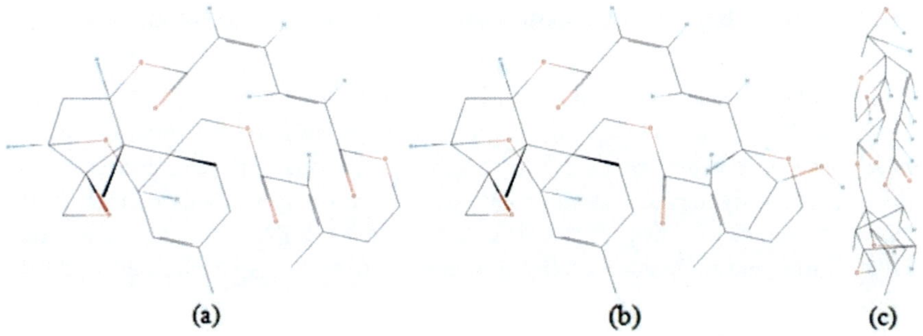

Fig. 5.22 Chemical structure of (**a**) verrucarin J (Pubchem), (**b**) roridin E (Pubchem), and (**c**) satratoxin H. (Adapted from Charlesy, Own work, Public Domain, https://commons.wikimedia.org/w/index.php?curid=2979803)

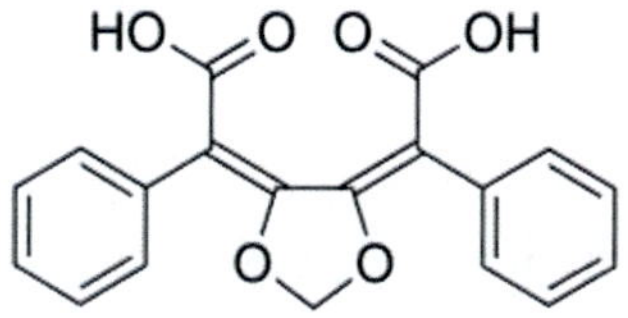

Fig. 5.23 Chemical structure of ustalic acid. (Adapted from Ed (Edgar181), Own work, Public Domain, https://commons.wikimedia.org/w/index.php?curid=25270655)

5.16.2 Ustalic Acid

Ustalic acid (Fig. 5.23) was isolated for the first time by Sano et al. in 2002 from the poisonous mushroom *Tricholoma ustale* (Sano et al. 2002). Its molecular weight is 338.3, while its molecular formula is $C_{19}H_{14}O_6$. The IUPAC name of ustalic acid is (2E)-2-[(5E)-5-[carboxy(phenyl)methylidene]-1,3-dioxolan-4-ylidene]-2-phenylacetic acid). Ustalic acid inhibits Na^+/K^+-ATPase, thus causing diarrhoea. Its intoxication in mice induces abdominal contraction and tremors (Sano et al. 2002).

5.16.3 Aziridine

Aziridine and its amino acid derivatives are produced by Angel's wing mushroom *Pleurocybella porrigens*. These toxins are considered vitamin D antagonists. *Pl. porrigens* toxin-induced encephalopathy exhibits demyelinating symptoms and showed mild toxicity to glial cells of mice (Sasaki et al. 2006).

5.17 Toxic Effects of Commonly Consumed Mushrooms

Some edible species of mushrooms also have adverse health effects. *Phellinus linteus* is known to increase the expression of TGFβ1, thus inducing prostatic hyperplasia in the experimental rats. Similarly, *Agaricus bisporus* is reported to have aromatic hydrazines; therefore, it acts as a carcinogen (Walton et al. 2001). However, *Agaricus silvaticus* has been observed to cause respiratory alterations, piloerection, and temporary apathy in male and female rats (Novaes et al. 2007). Osteolysin, the toxin isolated from edible oyster mushroom *Pleurotus* sp., is cytotoxic and can induce cardiac disorders like tachycardia, ischemia, and hypoxia in rats (Zuzek et al. 2006). The Japanese white mushroom Sugihiratake or Angel's wing mushroom (*Pleucybella porrigens*) can cause cryptogenic encephalopathy, usually in patients having kidney problems (Sasaki et al. 2006). In humans, the toxic effects of edible mushroom *Hapalopilus rutilans* include hypotonia, somnolence, visual disturbance, hepatic cytolysis, and kidney insufficiency (Lima et al. 2012).

5.18 Diagnosis of Mushroom Poisoning

Mushroom poisoning can be diagnosed based on the following criteria:

1. Correct identification of the mushroom consumed.
2. Proper case history of the patient.
3. Symptomatology.

Apart from these criteria, haematological test, detection of mushroom toxins in the serum and urine of the patient, and kidney and liver function tests should be performed for the baseline investigations (Barman et al. 2017).

5.19 Prevention and Control

The correct identification of the poisonous mushroom is the best way to avoid mushroom poisoning. The identifying features of poisonous mushrooms are the following:

1. Presence of volva.
2. Hollow stem.
3. Small brown mushrooms.
4. Presence of stem ring.
5. False morel mushrooms (blossom during fall).
6. Mushrooms with a pointed cap (Mishra et al. 2019).

Table 5.1 Summarizes specific treatment and supportive care strategies for mushroom poisoning (Vesconi et al. 1985; Köppel 1993; Barman et al. 2017)

S. No.	Treatment/supportive care	Recommendation
1.	GI intoxication by sequestering agents like medicinal/activated charcoal	Recommended dose: 50 g/kg/h for 4 days
2.	Gastric emptying	
3.	Haemodialysis	Recommended only for the patients with renal failure symptoms
4.	Biliary drainage followed by IV administration of octreotide	By gall bladder aspiration or removal of gall bladder for the protection of hepatocytes
5.	Antidotes or mushroom toxin inhibitors	Amatoxin inhibitors: Silibinin, intravenous penicillin G, silymarin, ceftazidime
6.	Antioxidants	*N*-acetylcysteine (NAC)
Supportive care		
1.	GI disturbances	Antiemetic and IV fluids
2.	Hallucination, agitation	Benzodiazepines
3.	Seizures	Benzodiazepines midazolam, phenytoin, sodium (as per epilepsy management protocol)
4.	Gyromitrin toxicity	Pyridoxine (70 mg to 5 g/kg/day) + anticonvulsant therapy, methylene blue (IV)
5.	Muscarine toxicity	Glycopyrrolate, anticholinergic therapy
6.	Rhabdomyolysis	Normal saline (20 ml to 1 l/kg/h)

Symptoms-based supportive care is the backbone of the treatment of mushroom poisoning (Table 5.1).

5.20 Future Prospects and Conclusion

Mushrooms are a great source of nutrients and are consumed worldwide. However, some poisonous mushrooms have hazardous health effects. Several studies have been conducted on fungal toxins. However, mushroom toxicology remains unexplored. Apart from this, structural elucidation and toxicological profiles of several mushroom toxins are poorly documented and require the proper attention of the scientific community. There is a need to raise public awareness about mushroom poisoning. Consumers should have detailed and up-to-date information on edible mushrooms (particularly identification) to avoid mushroom intoxication.

References

Barman B, Lynrah KG, Tiewsoh I (2017) Mushroom poisoning. Medicine Update-Section 13 Poisoning 2:538–541

Bedry R, Baudrimont I, Deffieux G, Creppy EE, Pomies JP, Ragnaud JM, Dupon M, Neau D, Gabinski C, De Witte S, Chapalain JC, Beylot J, Godeau P (2001) Wild-mushroom intoxication as a cause of rhabdomyolysis. N Engl J Med 345:798–802

Bogusz MJ, Maier RD, Schafer AT, Erkens M (1998) Honey with *Psilocybe* mushrooms: a revival of a very old preparation on the drug market? Int J Legal Med 111:147–150

Braun R, Dittmar W, Greeff U (1981) Considerations on the carcinogenicity of the mushroom poison gyromitrin and its metabolites. J Appl Toxicol 1:243–246

Catlow BJ, Song S, Paredes DA, Kirstein CL, Sanchez-Ramos J (2013) Effects of psilocybin on hippocampal neurogenesis and extinction of trace fear conditioning. Exp Brain Res 228:481–491

Chadeayne AR, Pham DNK, Reid BG, Golen JA, Manke DR (2020) Active metabolite of aeruginascin (4-Hydroxy-N,N,N-trimethyltryptamine): synthesis, structure, and serotonergic binding affinity. ACS Omega 5(27):16940–16943. https://doi.org/10.1021/acsomega.0c02208

Chang ST (2008) In: Cheung PCK (ed) Mushrooms as functional foods. Wiley, Hoboken, pp 1–34. https://doi.org/10.1002/9780470367285.ch1

Cho JT, Han JH (2016) A case of mushroom poisoning with *Russula subnigricans*: development of rhabdomyolysis, acute kidney injury, cardiogenic shock, and death. J Korean Med Sci 31(7): 1164–1167. https://doi.org/10.3346/jkms.2016.31.7.1164

Cohen-Addad C, Richard J-M, Guitel J-C (1987) Structures of an orellanine-trifluoroacetic acid comples: evidence of a very short OH...O hydrogen bond. Acta Crystallogr Sect C Cryst Struct Commun 43:504–507

Davoli P, Floriani M, Assisi F, Kob K, Sitta N (2016) Comment on "chemical and toxicological investigations of a previously unknown poisonous European mushroom *Tricholoma terreum*". Chem Eur J 22:5786–5788

De Bernardi M, Fronza G, Gianotti MP, Mellerio G, Vidari G, Vita-Finzi P (1983) Fungal metabolites XIII: new cytotoxic triterpene from *Hebeloma* species (Basidiomycetes). Tetrahedron Lett 24:1635–1638

Fujimoto H, Suzuki K, Hagiwara H, Yamazaki M (1986) New toxic metabolites from a mushroom, *Hebeloma vinosophyllum*. I. Structures of hebevinosides, I, II, III, IV, and V. Chem Pharm Bull 34:88–99

Fujimoto H, Takano Y, Yamazaki M (1992) Isolation, identification and pharmacological studies on three toxic metabolites from a mushroom. Chem Pharm Bull 40:869–872

Fushiya S, Sato S, Kanazawa T, Kusano G, Nozoe S (1990) Acromelic acid C, a new toxic constituent of *Clitocybe acromelalga*. An efficient isolation of acromelic acids. Tetrahedron Lett 31:3901–3904

Fushiya S, Kera Y, Nozoe S, Sato S (1992a) Isolation of acromelic acids D and E from *Clitocybe acromelalga*. Heterocycles 34:1277–1280

Fushiya S, Sato S, Nozoe S (1992b) L-Stizolobic acid and L-stizolobinic acid from *Clitocybe acromelalga,* precursors of acromelic acids. Phytochemistry 31:2337–2339

Fushiya S, Sato S, Kusano G, Nozoe S (1993) β-Cyano-L-alanine and N-(γ-L-glutamyl)-β-cyano-L-alanine, neurotoxic constituents of *Clitocybe acromelalga*. Phytochemistry 33:53–55

Garcia J, Costa VM, Carvalho A, Baptista P, de Pinho PG, de Lourdes BM, Carvalho F (2015) *Amanita phalloides* poisoning: mechanisms of toxicity and treatment. Food Chem Toxicol 86: 41–55

Gartz J (1987) Variation of the amount of alkaloids in fruit bodies of *Inocybe aeruginascens*. Planta Med 53(6):539–541. https://doi.org/10.1055/s-2006-962805

Gartz J (1997) Magic mushrooms around the world. LIS Publications, Los Angeles, p 27. ISBN 978-0-9653399-0-2

Gartz J (2008) Analysis of aeruginascin in fruit bodies of the mushroom *Inocybe aeruginascens*. Pharm Biol 27:141–144. https://doi.org/10.3109/13880208909053954

Hall AH, Spoerke DG, Rumack BH (1987) Mushroom poisoning: identification, diagnosis, and treatment. Pediatr Rev 8(10):291–298. https://doi.org/10.1542/pir.8-10-291

http://www.chemspider.com/Chemical-Structure.134192.html

http://www.chemspider.com/Chemical-Structure.86268.html

Jensen N, Gartz J, Laatsch H (2006) Aeruginascin, a trimethylammonium analogue of psilocybin from the hallucinogenic mushroom *Inocybe aeruginascens*. Planta Med 72(7):665–666. https://doi.org/10.1055/s-2006-931576

Karlson-Stiber C, Persson H (2003) Cytotoxic fungi–an overview. Toxicon 42:339–349

Kayano T, Kitamura N, Miyazaki S, Ichiyanagi T, Shimomura N, Shibuya I, Aimi T (2014) Gymnopilins, a product of a hallucinogenic mushroom, inhibit the nicotinic acetylcholine receptor. Toxicon 81:23–31

Kim HN, Do HH, Seo JS, Kim HY (2016) Two cases of incidental *Podostroma cornu-damae* poisoning. Clin Exp Emerg Med 3(3):186–189. https://doi.org/10.15441/ceem.15.028

Kirwan AP (1990) Magic mushroom' poisoning in a dog. Vet Record 26(6):149

Konno K, Shirahama H, Matsumoto T (1983) Isolation and structure of acromelic acid A and B. New kainoids of *Clitocybe acromelalga*. Tetrahedron Lett 24:939–942

Köppel C (1993) Clinical symptomatology and management of mushroom poisoning. Toxicon 31(12):1513–1540. https://doi.org/10.1016/0041-0101(93)90337-i

Leung AY, Paul AG (1968) Baeocystin and norbaeocystin: new analogs of psilocybin from *Psilocybe baeocystis*. J Pharm Sci 57(10):1667–1671. https://doi.org/10.1002/jps.2600571007

Lima ADL, Costa Fortes R, Novaes MRCG, Percário S (2012) Poisonous mushrooms; a review of the most common intoxications. Nutr Hosp 27(2):402–408. https://doi.org/10.3305/nh.2012.27.2.5328

List PH, Luft P (1967) Gyromitrin, das Gift der Fruhjahrslorchel *Gyromitra (Helvella) esculenta* Fr. Tetrahedron Lett 8:1893–1894

Loranger A, Tuchweber B, Gicquaud C, St-Pierre S, Cote MG (1985) Toxicity of peptides of *Amanita virosa* mushrooms in mice. Toxicol Sci 5:1144–1152

Luo H, Hallen-Adams HE, Scott-Craig JS, Walton JD (2012) Ribosomally encoded cyclic peptide toxins from mushrooms. Fungal Genet Biol 49:123–129

Matsuura M, Saikawa Y, Inui K, Nakae K, Igarashi M, Hashimoto K, Nakata M (2009) Identification of the toxic trigger in mushroom poisoning. Nat Chem Biol 5:465–467

McMorris TC, Anchel M (1965) Fungal metabolites. The structures of the novel sesquiterpenoids illudin-S and -M. J Am Chem Society 87:1594–1600

Miles PG, Chang ST (2004) Mushrooms: nutritional value, medicinal effect, and environmental impact, 2nd edn. CRC Press, Boca Raton, pp 1–26

Minami T, Matsumura S, Nishizawa M, Sasaguri Y, Hamanaka N, Ito S (2004) Acute and late effects on induction of allodynia by acromelic acid, a mushroom poison related structurally to kainic acid. Br J Pharmacol 142:679–688

Mishra KK, Mishra PK, Stanley J (2019) Mushroom poisoning. Indian Farmers Digest 52(6):32–36

Nishio A, Kitamura N, Tanaka S, Miyazaki S, Ichiyanagi T, Shimomura N, Shibuya I, Aimi T (2012) Multiple effects of gymnopilin on circulatory system of the rat. Biol Pharm Bull 35(8): 1300–1305. https://doi.org/10.1248/bpb.b12-00183

Novaes MRCG, Novaes LCG, Melo AL, Recova VI (2007) Effects of *Agaricus sylvaticus* in immune and hematopoietic system of rats with Walker 256 ascitic tumor. Rev Bras Nutr Clín 22(2):116–120

Nozoe S, Koike Y, Kusano G, Seto H (1983) Structure of gymnopilin, a bitter principle of an hallucinogenic mushroom, *Gymnopilus spectabilis*. Tetrahedron Lett 24(16):1735–1736. https://doi.org/10.1016/S0040-4039(00)81757-8

Pajouhesh H, Lenz GR (2005) Medicinal chemical properties of successful central nervous system drugs. NeuroRx 2(4):541–553. https://doi.org/10.1602/neurorx.2.4.541

Patocka J, Wu R, Nepovimova E, Valis M, Wu W, Kuca K (2021) Chemistry and toxicology of major bioactive substances in Inocybe mushrooms. Int J Mol Sci 22(4):2218. https://doi.org/10.3390/ijms22042218

Piqueras J (2000) Aminita muscaria & Amanita pantherina and others. International Programme on Chemical Safety. Poisons Information Monograph (Group monograph) G026 Fungi. https://inchem.org/documents/pims/fungi/pimg026.htm#PartTitle:14.%20AUTHOR(S),%20

REVIEWER(S)%20DATA%20(INCLUDING%20EACH%20UPDATING),%20COMPLETE%20ADDRESSE

Puig MV, Artigas F, Celada P (2005) Modulation of the activity of pyramidal neurons in rat prefrontal cortex by raphe stimulation *in vivo*: involvement of serotonin and GABA. Cereb Cortex 15(1):1–14. https://doi.org/10.1093/cercor/bhh104

Puschner B (2018) Mushroom toxins. In: Gupta RC (ed) Veterinary toxicology, 3rd edn. Academic Press, Boston, pp 955–966. ISBN 9780128114100. https://doi.org/10.1016/B978-0-12-811410-0.00067-2

Pyysalo H, Niskanen A (1977) On the occurrence of N-methyl-N-formylhydrazones in fresh and processed false morel, *Gyromitra esculenta*. J Agric Food Chem 25:644–647

Sano Y, Sayama K, Arimoto Y, Inakuma T, Kobayashi K, Koshino H, Kawagishi H (2002) Ustalic acid as a toxin and related compounds from the mushroom *Tricholoma ustale*. Chem Commun (Camb) 13:1384–1385. https://doi.org/10.1039/b202607d

Sasaki H, Akiyama H, Yoshida Y, Kondo K, Amakura Y, Kasahara Y, Maitani T (2006) Sugihiratake mushroom (angel's wing mushroom)-induced cryptogenic encephalopathy may involve vitamin D analogues. Biol Pharm Bull 29(12):2514–2518. https://doi.org/10.1248/bpb.29.251

Schwarcz R, Hökfelt T, Fuxe K, Jonsson G, Goldstein M, Terenius L (1979) Ibotenic acid-induced neuronal degeneration: a morphological and neurochemical study. Exp Brain Res 37:199–216. https://doi.org/10.1007/BF00237708

Sherwood AM, Halberstadt AL, Klein AK, McCorvy JD, Kaylo KW, Kargbo RB, Meisenheimer P (2020) Synthesis and biological evaluation of tryptamines found in hallucinogenic mushrooms: norbaeocystin, baeocystin, norpsilocin, and aeruginascin. J Nat Prod 83(2):461–467

Shi GQ, Huang WL, Zhang J, Zhao H, Shen T, Fontaine RE, Yang L, Zhao S, Lu BL, Wang YB, Ma L, Li ZX, Gao Y, Yang ZL, Zeng G (2012) Clusters of sudden unexplained death associated with the mushroom, *Trogia venenata* in rural Yunnan Province, China. PLoS One 7:e35894

Taguchi T, Tomotoshi K, Mizumura K (2009) Excitatory actions of mushroom poison (acromelic acid) on unmyelinated muscular afferents in the rat. Neurosci Lett 456:69–73

Tanaka M, Hashimoto K, Okuno T, Shirahama H (1993) Neurotoxic oligoisoprenoids of the hallucinogenic mushroom, *Gymnopilus spectabilis*. Phytochemistry 34(3):661–664

Vesconi S, Langer M, Iapichino G, Costantino D, Busi C, Fiume L (1985) Therapy of cytotoxic mushroom intoxication. Crit Care Med 13(5):402–406. https://doi.org/10.1097/00003246-198505000-00007

Vetter J (1998) Toxins of *Amanita phalloides*. Toxicon 36:13–24

Voynova M, Shkondrov A, Kondeva-Burdina M, Krasteva I (2020) Toxicological and pharmacological profile of Amanita muscaria (L.) Lam. – a new rising opportunity for biomedicine. Pharmacia 67:317–323. https://doi.org/10.3897/pharmacia.67.e56112

Walton K, Coombs MM, Walker R, Ioannides C (2001) The metabolism and bioactivation of agaritine and of other mushroom hydrazines by whole mushroom homogenate and by mushroom tyrosinase. Toxicology 161(3):165–177. https://doi.org/10.1016/s0300-483x(00)00430-3

White J, Weinstein SA, De Haro L, Bédry R, Schaper A, Rumack BH, Zilker T (2019) Mushroom poisoning: a proposed new clinical classification. Toxicon 157:53–65. https://doi.org/10.1016/j.toxicon.2018.11.007

Wieland T, Luben G, Ottenheym H, Faesel J, De Vries JX, Prox A, Schmid J (1968) The discovery, isolation, elucidation of structure, and synthesis of antamanide. Angew Chem Int Ed Engl 7:204–208

Yamaura YFM, Fukuhara M, Takabatake E, Ito N, Hashimoto T (1986) Hepatotoxic action of a poisonous mushroom, *Amanita abrupta* in mice and its toxic component. Toxicology 38:161–173

Yin X, Yang A-A, Gao J-M (2019) Mushroom toxins: chemistry and toxicology. J Agric Food Chem 67:5053–5071. https://doi.org/10.1021/acs.jafc.9b00414

Yoshikawa K, Kuroboshi M, Arihara S, Miura N, Tujimura N, Sakamoto K (2002) New triterpenoids from *Tricholoma saponaceum*. Chem Pharm Bull 50(12):1603–1606. https://doi.org/10.1248/cpb.50.1603

Zhou ZY, Shi GQ, Fontaine RE, Wei K, Feng T, Wang F, Wang GQ, Qu Y, Li ZH, Dong ZJ, Zhu HJ, Yang ZL, Zeng G, Liu JK (2012) Evidence for the natural toxins from the mushroom *Trogia venenata* as a cause of sudden unexpected death in Yunnan province, China. Angew Chem Int Ed 51:2368–2370

Zuzek MC, Macek P, Sepcić K, Cestnik V, Frangez R (2006) Toxic and lethal effects of ostreolysin, a cytolytic protein from edible oyster mushroom (*Pleurotus ostreatus*), in rodents. Toxicon 48(3):264–271. https://doi.org/10.1016/j.toxicon.2006.05.011

Chapter 6
Mycotoxins Co-occurrence Poisoning

6.1 Introduction

Mycotoxins are usually produced in pre- or post-harvest periods when crops are in stressed conditions and become susceptible to the growth of various moulds. The prevalence of natural occurrence of multi-mycotoxins is attributable to an increase in global warming (Tolosa et al. 2021). The climatic conditions contribute significantly to post-harvest mycotoxin contamination (Abbas 2019). The co-infection of multiple mycotoxigenic fungi and simultaneous production of more than one mycotoxin from a single fungal species leads to the production of a mycotoxin cocktail in the crops. The major mycotoxins like AFs, FBs, DON, and ZEN are produced in similar climatic conditions, thus co-contaminating the crops (Paterson and Lima 2010). Moreover, the same geographical location, similar harvesting, storing, and processing methods also contribute significantly the co-occurrence of the mycotoxins. Apart from this, the multi-commodity composition of food and feed is another reason for mycotoxin co-occurrence. The co-contamination of mycotoxins is frequently found in many agricultural commodities (Ingenbleek et al. 2019; Gruber-Dorninger et al. 2019), forage crops, and silages (Gallo et al. 2015). Moreover, agriculture-based products have been regularly tested for the presence of only a few major mycotoxins, like aflatoxins, ochratoxins, zearalenone. Despite the risk associated with toxicological interactions of mycotoxins, there is hardly any regulation on their combined occurrence globally (Grenier and Oswald 2011; Smith et al. 2016).

K. Singh, A. Kumari, *Mycotoxins and Mycotoxicoses*,
https://doi.org/10.1007/978-981-19-2370-8_6

6.2 Multi-Mycotoxins in Food

The co-occurrence of DON, NIV, ZEN, OTA, T-2, and HT-2 mycotoxins has been predominantly reported worldwide (Palumbo et al. 2020). However, there are only a few reports on the co-occurrence of AFs, *Alternaria* toxins, and ENNs. In 1987, Abramson et al. reported the natural occurrence of deoxynivalenol (DON), diacetoxyscirpenol, HT-2, and T-2 toxins in 53 samples of wheat from southern Manitoba (Abramson et al. 1987). The natural occurrence of nivalenol (NIV), zearalenone (ZEN), and deoxynivalenol (DON) was found in nine samples of wheat and one sample each of corn and rye from Canada (Tanaka et al. 1988).

Similarly, the co-contamination of aflatoxins, fumonisins, nivalenol, zearalenone, and deoxynivalenol was detected in 16 corn samples from Indonesia (Ali et al. 1998). The natural occurrence of fumonisin B_1, fumonisin B_2, aflatoxin B_1, aflatoxin G_1, and aflatoxin G_2 was found in 15 pre-harvest maize samples collected from Nigeria (Bankole and Mabekoje 2004). In 2006, Abbas et al. analysed the effect of drought-like conditions and found that the high heat favoured the production of aflatoxins and fumonisins in harvested maize (Abbas et al. 2006). Also, 27 samples of processed cereal-based baby food and milk powder collected from a Portuguese market were found contaminated with aflatoxin M_1 and ochratoxin A (Alvito et al. 2010). By applying the HPLC-MS/MS method, de Souza et al. (2013) detected the presence of fumonisins B_1, B_2, ZEN, aflatoxins B_1, B_2, G_1, G_2, agroclavine, hydrolysed fumonisin B, chanoclavine, NIV, MON, BEA, kojic acid, and ENN A, A_1, B, B_1 in maize and poultry feed (de Souza et al. 2013). Human food and dietary exposure consisting of rice, wheat flour, and corn at a high risk (of liver and oesophageal cancers) area of China were found to be co-contaminated with AFB_1 and FB_1 (Sun et al. 2011). Similarly, Kamala et al. (2015) reported natural occurrences of FB1, FB2, AFB1, AFB2, AFG1, AFG2, OTA, DON, HT-2 toxin, and ZEN in maize kernels collected from three agro-ecological zones of Tanzania (Kamala et al. 2015). Likewise, the presence of AFB1, AFB2, AFG1, AFG2, α-ZEN, and CPA was detected in the samples of dried beef (Dada et al. 2020).

6.3 Mycotoxin Co-occurrence in Animal Feeds

Co-contamination of mycotoxins in the feeding stuff affects animals' health even at a very low concentration. The animal feed is composed of several ingredients like cereals, crops, by-products of human food, and vegetable residues, thereby favouring the growth of multiple mycotoxigenic fungi. Several countries monitor the presence and co-occurrence of many major mycotoxins like AFs, ZEN, DON, and FBs in animal feeds. However, the co-contamination of emerging mycotoxins is poorly managed, especially in developing countries. Natural co-occurrence of mycotoxins in composite feed and animal feeding stuff has been reported worldwide (Tolosa et al. 2021). In 2012, Rodrigues and Nachrer evaluated 6000 samples of

Table 6.1 Mycotoxin co-occurrence in the animal feed

S. No.	Type of feed	Mycotoxin combination	Reference
1.	Feedstuffs containing wheat, corn, corn gluten meal, rice, soybean meal, silage, barley, dried distillers grains (DDGS)	DON + FUM	BIOMIN (2016)
2.	Wheat (unprocessed; wheat flour)	NIV + DON + DON derivatives + T-2 and HT-2 + ZEN + emerging toxins	Tima et al. (2016), Stanciu et al. (2017)
3.	Spanish barley	AF + ZEN + OTA	Ibáñez-Vea et al. (2012)
4.	Feeding stuff and complex compound feeds	ZEN + DON + OTA + FBs + *Alternaria* toxins + BEA + ergot alkaloids + ENNs	Zachariasova et al. (2014)
5.	Maize, complete feed samples, barley, oat	DON + NIV + ZEN + T-2 + AFs + FBs + OTA + ENNs + BEA	Tima et al. (2016), Kosicki et al. (2016), Tolosa et al. (2019)
6.	Broiler feed	ZEN + DON + FUMs	www.dairyglobal.net
7.	Feed for cow, sheep, equine, goat and poultry	AFB_1 + ZEN + OTA + DON + T2 + FB_1 + FB_2	Almeida et al. (2013), Křížová et al. (2021)
8.	Pet food	BEA + ENNs	Castaldo et al. (2019)
9.	Soy animal feed	FBs + AFs + DON	Gutleb et al. (2015)

feedstuff and 81% of them were found contaminated with two or more mycotoxins. Similarly, about 64% of samples of animal feed were observed to be co-contaminated with ZEN and DON or FBs and ZEN or FBs and DON in a global survey conducted by BIOMIN Company (Tolosa et al. 2021). The most frequently detected dual contaminations of mycotoxins in animal feed from different parts of the world are the following:

1. Wheat-based feed contaminated with ZEN and DON in South and Southeast Asia and Oceania
2. AFB_1 and FBs in finished feed/maize from Sub-Saharan Africa
3. OTA and AFB_1 in finished feed from South Asia (Gruber-Dorninger et al. 2019)

The study conducted on multi-mycotoxin in animal (pig fattening) feeds reported DON + ZEN as the most frequently co-occurring mycotoxins (Almeida et al. 2011). But, in 2016, the co-occurrence of DON + FUM was predominantly detected in the feeding stuff (BIOMIN 2016). Nevertheless, the highest percentage (80%) of mycotoxin co-contamination was recorded in poultry feed samples from Asia (BIOMIN 2017). The overview of mycotoxin co-occurrence is given in Table 6.1.

6.4 Mycotoxin Co-occurrence Toxicity

The synergistic, additive, and antagonistic interactions among mycotoxins often lead to severe health problems in humans as well as in animals (Smith et al. 2016). The toxicity caused by mycotoxin co-contamination is usually unpredictable.

6.4.1 Mycotoxin Co-occurrence Toxicity in Humans

The human diet is composed of many agricultural and animal products. The ubiquitous presence of mycotoxins in unprocessed/raw and processed food products has been observed to cause mycotoxin co-occurrence poisoning. The level of mycotoxins can be measured by their biomarkers present in the patient's urine (Solfrizzo et al. 2011). The persons exposed to environments contaminated with mycotoxigenic fungi exhibited the presence of trichothecenes, ochratoxin, and aflatoxin in their tissues, urine, and sputum (Hooper et al. 2009). The mycotoxins FB1, DON15G, ZEN α and β zearalenol, DON, and OTA were found in the urine of farmers in the Transkei region of South Africa, known for high incidences of oesophageal cancer (Shephard et al. 2013). However, DON-3-glucuronide, deoxynivalenol, enniatin B, T-2 toxin, zearalenone-14-*O*-glucuronide, and dihydrocitrinone were detected in the urine samples of a German population and AFM_1, dihydrocitrinone, and OTA in the populations of Bangladesh and Haiti (Gerding et al. 2015). The urine samples of the patients suffering from chronic fatigue syndrome (CFS) were found positive for OTA and macrocyclic trichothecenes (Brewer et al. 2013). The presence of enniatin B, 2′R-OTA, and OTA was recorded in the blood and serum samples of the HIV-positive population in Cameroon (Osteresch et al. 2017).

6.4.2 Mycotoxin Co-occurrence Toxicity in Animals

In animals, the young ones are more susceptible to multiple mycotoxin toxicity than adults. The animal feeds consist of a variety of feeding stuff including cereal-based products, vegetables, and by-products of human food, thus exposing them to a mixture of multiple mycotoxins. The toxicological interactions of mycotoxins have either synergistic or additive effects on animals (Grenier and Oswald 2011). The co-contamination of OTA, penicillic acid, and FB_1 was found to induce mycotoxic nephropathy in Bulgarian pig and poultry farms (Stoev et al. 2010). The co-occurrence of FB_1 and AFB_1 is a matter of concern as they promote tumour formation in the liver (Grenier and Oswald 2011). A dual combination of AFs and DON has been reported to cause impaired immunity and hepatotoxicity (Weaver et al. 2013) whilst DON and FBs exhibited a synergistic inhibitory effect on weight gain and an additive inhibitory effect on food intake in pigs (Grenier and Oswald 2011). However, the binary combinations of DON + AFB_1 and ZEN + DON have been found to affect ileal digestibility in swine and immune functions in both mice and swine, respectively (Dabrowski et al. 2016; Holanda et al. 2020). The antagonistic effect of AFB_1 + FB_1 induces carcinogenicity, whereas their synergistic interaction results in immunotoxicity (Alassane-Kpembi et al. 2017). Likewise, a synergistic inhibitory effect was observed on the proliferation of swine intestinal cells when exposed to deoxynivalenol, ZEN, and FB_1 (Alassane-Kpembi et al. 2017).

6.5 Future Prospects and Conclusion

The occurrence of multiple mycotoxins in agricultural commodities has been widely reported all over the world. However, the knowledge about their co-exposure to human and animal health is limited. In addition to that, the risk associated with toxicological interactions of emerging mycotoxins and their modified forms is still unexplored. Therefore, more studies ought to be done in this direction. Moreover, detection of mycotoxin combinations and their interactions should be undertaken using highly sensitive methods. The implementation of mycotoxin decontamination strategies, preventive measures (like crop rotation), and refined assessment can reduce the possibility of mycotoxin co-occurrence and their intoxications. Furthermore, the natural co-contamination of mycotoxins may impose high toxicity. Thus, studies on multiple mycotoxins should be conducted in realistic patterns.

References

Abbas M (2019) Co-occurrence of mycotoxins and its detoxification strategies. In: Mycotoxins-impact and management strategies. IntechOpen. https://doi.org/10.5772/intechopen.76562

Abbas HK, Cartwright RD, Xie W, Shier WT (2006) Aflatoxin and fumonisin contamination of corn (maize, Zea mays) hybrids in Arkansas. Crop Prot 25(1):1–9

Abramson D, Clear RM, Nowicki TW (1987) *Fusarium* species and trichothecene mycotoxins in suspect samples of 1985 Manitoba wheat. Can J Plant Sci 67(3):611–619. https://doi.org/10.4141/cjps87-087

Alassane-Kpembi I, Schatzmayr G, Taranu I, Marin D, Puel O, Oswald IP (2017) Mycotoxins co-contamination: methodological aspects and biological relevance of combined toxicity studies. Crit Rev Food Sci Nutr 57:3489–3507

Ali N, Sardjono YA, Yoshizawa T (1998) Natural co-occurrence of aflatoxins and *Fusarium* mycotoxins (fumonisins, deoxynivalenol, nivalenol and zearalenone) in corn from Indonesia. Food Addit Contam 15:377–384

Almeida I, Martins HM, Santos S, Costa JM, Bernardo F (2011) Co-occurrence of mycotoxins in swine feed produced in Portugal. Mycotox Res 27:177–181

Almeida I, Guerra M, Martins HM, Costa JM, Bernardo F (2013) Aflatoxin B1 and zearalenone in dairy feeds in Portugal, 2009–2011. Mycotoxin Res 29:131–133. https://doi.org/10.1007/s12550-013-0164-0

Alvito PC, Sizoo EA, Almeida CMM, van Egmond HP (2010) Occurrence of aflatoxins and ochratoxin A in baby foods in Portugal. Food Anal Methods 3:22–30

Bankole SA, Mabekoje OO (2004) Occurrence of aflatoxins and fumonisins in preharvest maize from south-western Nigeria. Food Addit Contam 21:251–255

BIOMIN (2016) Science & Solutions. BIOMIN Holding GmbH, Herzogenburg

BIOMIN (2017) Science & Solutions. BIOMIN Holding GmbH, Herzogenburg

Brewer JH, Thrasher JD, Straus DC, Madison RA, Hooper D (2013) Detection of mycotoxins in patients with chronic fatigue syndrome. Toxins (Basel) 5(4):605–617. https://doi.org/10.3390/toxins5040605

Castaldo L, Graziani G, Gaspari A, Izzo L, Tolosa J, Rodríguez-Carrasco Y, Ritieni A (2019) Target analysis and retrospective screening of multiple mycotoxins in pet food using UHPLC-Q-orbitrap HRMS. Toxins 11:434. https://doi.org/10.3390/toxins11080434

Dabrowski M, Obremski K, Gaje Cka M, Gaje Cki MT, Zielonka Ł (2016) Changes in the subpopulations of porcine peripheral blood lymphocytes induced by exposure to low doses of zearalenone (ZEN) and deoxynivalenol (DON). Molecules 21:557

Dada TA, Ekwomadu TI, Mwanza M (2020) Multi mycotoxin determination in dried beef using liquid chromatography coupled with Triple Quadrupole Mass Spectrometry (LC-MS/MS). Toxins (Basel) 12(6):357. https://doi.org/10.3390/toxins12060357

Gallo A, Giuberti G, Frisvad JC, Bertuzzi T, Nielsen KF (2015) Review on mycotoxin issues in ruminants: occurrence in forages, effects of mycotoxin ingestion on health status and animal performance and practical strategies to counteract their negative effects. Toxins 7(8):3057–3111

Gerding J, Ali N, Schwartzbord J, Cramer B, Brown DL, Degen GH, Humpf HU (2015) A comparative study of the human urinary mycotoxin excretion patterns in Bangladesh, Germany, and Haiti using a rapid and sensitive LC-MS/MS approach. Mycotox Res 31:127–136. https://doi.org/10.1007/s12550-015-0223-9

Grenier B, Oswald IP (2011) Mycotoxin co-contamination of food and feed: meta-analysis of publications describing toxicological interactions. World Mycotoxin J 4:285–313

Gruber-Dorninger C, Jenkins T, Schatzmayr G (2019) Global mycotoxin occurrence in feed: a ten-year survey. Toxins 11:375. https://doi.org/10.3390/toxins11070375

Gutleb AC, Caloni F, Giraud F, Cortinovis C, Pizzo F, Hoffmann L, Bohn T, Pasquali M (2015) Detection of multiple mycotoxin occurrences in soy animal feed by traditional mycological identification combined with molecular species identification. Toxicol Rep 2:275–279. https://doi.org/10.1016/j.toxrep.2015.01.006

Holanda DM, Yiannikouris A, Kim SW (2020) Investigation of the efficacy of a postbiotic yeast cell wall-based blend on newly-weaned pigs under a dietary challenge of multiple mycotoxins with emphasis on deoxynivalenol. Toxins 12:504

Hooper DG, Bolton VE, Guilford FT, Straus DC (2009) Mycotoxin detection in human samples from patients exposed to environmental molds. Int J Mol Sci 10(4):1465–1475. https://doi.org/10.3390/ijms10041465. PMID: 19468319; PMCID: PMC2680627

Ibáñez-Vea M, Lizarraga E, González-Peñas E, López de Cerain A (2012) Cooccurrence of type-A and type-B trichothecenes in barley from a northern region of Spain. Food Control 25:81–88. https://doi.org/10.1016/j.foodcont.2011.10.028

Ingenbleek L, Sulyok M, Adegboye A, Hossou SE, Koné AZ, Oyedele AD, Kisito KJ, Koreissi CS, Dembélé Y, Eyangoh S, Verger P, Leblanc JC (2019) Regional Sub-Saharan Africa total diet study in Benin, Cameroon, Mali and Nigeria reveals the presence of 164 mycotoxins and other secondary metabolites in foods. Toxins 11(1):54

Kamala A, Ortiz J, Kimanya M, Haesaert G, Donoso S, Tiisekwa B, De Meulenaer B (2015) Multiple mycotoxin co-occurrence in maize grown in three agro-ecological zones of Tanzania. Food Control 54:208–215. https://doi.org/10.1016/j.foodcont.2015.02.002

Kosicki R, Błajet-Kosicka A, Grajewski J, Twarużek M (2016) Multiannual mycotoxin survey in feed materials and feeding stuffs. Anim Feed Sci Technol 215:165–180. https://doi.org/10.1016/j.anifeedsci.2016.03.012

Křížová L, Dadáková K, Dvořáčková M, Kašparovský T (2021) Feedborne mycotoxins beauvericin and enniatins and livestock animals. Toxins (Basel) 13(1):32. https://doi.org/10.3390/toxins13010032

Osteresch B, Viegas S, Cramer B, Humpf H-U (2017) Multi-mycotoxin analysis using dried blood spots and dried serum spots. Anal Bioanal Chem 409:3369–3382. https://doi.org/10.1007/s00216-017-0279-9

Palumbo R, Crisci A, Venâncio A, Cortiñas Abrahantes J, Dorne JL, Battilani P, Toscano P (2020) Occurrence and co-occurrence of mycotoxins in cereal-based feed and food. Microorganisms 8 (1):74

Paterson RRM, Lima N (2010) How will climate change affect mycotoxins in food? Food Res Int 43:1902–1914

Shephard GS, Burger HM, Gambacorta L, Gong YY, Krska R, Rheeder JP, Solfrizzo M, Srey C, Sulyok M, Visconti A, Warth B (2013) Multiple mycotoxin exposure determined by urinary biomarkers in rural subsistence farmers in the former Transkei, South Africa. Food Chem Toxicol 62:217–225

Smith MC, Madec S, Coton E, Hymery N (2016) Natural co-occurrence of mycotoxins in foods and feeds and their in vitro combined toxicological effects. Toxins 8(4):94

Solfrizzo M, Gambacorta L, Lattanzio VM, Powers S, Visconti A (2011) Simultaneous LC-MS/MS determination of aflatoxin M1, ochratoxin A, deoxynivalenol, de-epoxydeoxynivalenol, α and β-zearalenols and fumonisin B1 in urine as a multi-biomarker method to assess exposure to mycotoxins. Anal Bioanal Chem 401:2831–2841. https://doi.org/10.1007/s00216-011-5354-z

de Souza MLM, Sulyok M, Freitas-Silva O, Costa SS, Brabet C, Machinski MJ, Sekiyama BL, Vargas EA, Krska R, Schuhmacher R (2013) Cooccurrence of mycotoxins in maize and poultry feeds from Brazil by Liquid Chromatography/Tandem Mass Spectrometry. Sci World J 9: 427369. https://doi.org/10.1155/2013/427369

Stanciu O, Juan C, Miere D, Loghin F, Mañes J (2017) Occurrence and cooccurrence of Fusarium mycotoxins in wheat grains and wheat flour from Romania. Food Control 73:147–155. https://doi.org/10.1016/j.foodcont.2016.07.042

Stoev SD, Dutton MF, Njobeh PB, Mosonik JS, Steenkamp PA (2010) Mycotoxic nephropathy in Bulgarian pigs and chickens: complex aetiology and similarity to Balkan Endemic Nephropathy. Food Addit Contam A Chem Anal Cont Expos Risk Assess 27:72–88

Sun G, Wang S, Hu X, Su J, Zhang Y, Xie Y, Zhang H, Tang L, Wang JS (2011) Co-contamination of aflatoxin B1 and fumonisin B1 in food and human dietary exposure in three areas of China. Food Additiv Contam A 28(4):461–470. https://doi.org/10.1080/19440049.2010.544678

Tanaka T, Hasegawa A, Yamamoto S, Lee US, Sugiura Y, Ueno Y (1988) A case report on a minor contamination of nivalenol in cereals harvested in Canada. Mycopathologia 101:157–160. https://doi.org/10.1007/BF00437032

Tima H, Brückner A, Mohácsi-Farkas C, Kiskó G (2016) *Fusarium* mycotoxins in cereals harvested from Hungarian fields. Food Addit Contam B Surveil 9(2):127–131. https://doi.org/10.1080/19393210.2016.1151948

Tolosa J, Rodríguez-Carrasco Y, Ferrer E, Mänes J (2019) Identification and quantification of enniatins and beauvericin in animal feeds and their ingredients by LC-QTRAP/MS/MS. Metabolites 9:33. https://doi.org/10.3390/metabo902003

Tolosa J, Rodríguez-Carrasco Y, Ruiz MJ, Vila-Donat P (2021) Multi-mycotoxin occurrence in feed, metabolism and carry-over to animal-derived food products: a review. Food Chem Toxicol 158:112661

Weaver AC, See MT, Hansen JA, Kim YB, De Souza ALP, Middleton TF, Kim SW (2013) The use of feed additives to reduce the effects of aflatoxin and deoxynivalenol on pig growth, organ health and immune status during chronic exposure. Toxins 5:1261–1281

Zachariasova M, Dzuman Z, Veprikova Z, Hajkova K, Jiru M, Vaclavikova M, Zachariasova A, Pospichalova M, Florian M, Hajslova J (2014) Occurrence of multiple mycotoxins in European feeding stuffs, assessment of dietary intake by farm animals. Anim Feed Sci Technol 193:124–140. https://doi.org/10.1016/j.anifeedsci.2014.02.007

Chapter 7
Masked and New Mycotoxins

7.1 Introduction

Native mycotoxins convert into their derivatives when subjected to plant metabolism or processing of contaminated agricultural commodities. These derivatives co-exist with their parent compounds and usually could not be detected by regular screening of food and feedstuff. They are known as **masked mycotoxins** (Galaverna et al. 2009). In 2014, Rychlik et al. suggested the use of the term **modified mycotoxins** in place of masked mycotoxins and described them in the following hierarchal levels:

1. **The highest level** separates unmodified and free forms from the chemically modified and matrix-associated forms of mycotoxins.
2. **Matrix-associated forms** are further differentiated into complexes and covalently bound forms.
3. **Modified forms** include biologically modified and chemically modified forms.
4. **Chemically modified forms** further differentiated into non-thermally formed and thermally formed forms.
5. **Biologically modified forms**, however, are categorized into **phase I metabolites (Functionalized)** and **phase II metabolites (Conjugated)**. Conjugated forms have three types:

    (a) **Conjugated by plants (masked)** Ex-DON-3 glucoside
    (b) **Conjugated by animals** Ex-HT-2-3/4 glucuronide
    (c) **Conjugated by fungi** Ex-Deepoxy-DON (DOM-1) and ZEN-14-sulphate

The first report of modified mycotoxins was published in 1967 by Masri et al. They reported the presence of AFM_1, a metabolite of aflatoxin B_1 in the milk (Masri et al. 1967). However, the first mycotoxin detected as masked mycotoxin was zearalenone-14-glucoside (www.dairyglobal.net/health, Diary Global n.d.). Later in the 1980s, several cases of mycotoxicity were reported in the animals; nonetheless, a very low concentration of mycotoxins was detected in their feed (Berthiller et al. 2015). At the same time, a less toxic metabolite of DON was found in the

K. Singh, A. Kumari, *Mycotoxins and Mycotoxicoses*,
https://doi.org/10.1007/978-981-19-2370-8_7

Ontario winter wheat and field corn with *Fusarium graminearum* infestation (EFSA 2007). In the early 1990s, the structure of a metabolite of 4-deoxynivalenol was elucidated (Sewald et al. 1992). However, the natural occurrence of a masked mycotoxin DON-3-Glc was first time reported in maize and wheat in 2005 (Berthiller et al. 2005). Thereafter, the presence of modified metabolites of ZEN, fumonisins, T-2 and HT-2 toxins has also been recorded (De Boevre et al. 2012; Falavigna et al. 2012; Veprikova et al. 2012).

7.2 Generation of Masked Mycotoxins in Plants

Masked mycotoxins are the products of chemical transformation catalysed by detoxification enzymes of plants (Berthiller et al. 2013). Plants use hydrolysis, oxidation, and conjugation reactions for the modification of native mycotoxins. The hydrolysis reaction (phase I) is catalysed by amidases and esterases while the cytochrome P-450 system catalyses the oxidation reaction (Coleman et al. 1997). The toxicity of phase I metabolites is substantially high but the products of conjugation reaction (phase II) are either non-toxic or have less toxicity than their native forms. In conjugation reactions, glucosyl residues can bind with carboxy, amino, hydroxyl, and thiol functional groups. The binding of malonyl residues, however, takes place with amino and hydroxyl groups only. All the phase II reactions are catalysed by respective transferases (Berthiller et al. 2013). The glycosylation and glutathione conjugation detoxify mycotoxins by converting them into masked mycotoxins. Stress conditions like herbicide treatment induce the production of masked mycotoxins as well (Cummins et al. 2011).

7.3 Generation of Masked Mycotoxins During Processing of Food

Mycotoxin-contaminated fruits and cereals when undergoing food processing transform into masked mycotoxins. These derivatives are not detected by routine methods used for the monitoring of mycotoxins. The microbial enzymes used in food processing convert native mycotoxins into masked ones. However, processed food products like pickles, jams, wines, fermented sausages, and beers have not been screened extensively for the occurrence of masked mycotoxins (Berthiller et al. 2013). Presence of fusaric acid methylamide and bound FBs were found in the infected vegetables and cereal-based food while the natural occurrence of zearalenone-14-β-D-glucopyranoside (Z14G) and deoxynivalenol-3-glucoside (D3G) was reported in barley, maize, and wheat-based products (Berthiller et al. 2013). The modified forms of the native mycotoxins in food and food products are illustrated in Table 7.1.

Table 7.1 Masked mycotoxins in food and cereal-based products

S. No.	Food/food products	Native mycotoxin	Masked mycotoxin	Reference/s
1.	Corn flakes; heat-processed corn food; gluten-free food products	Fumonisins (FB_1)	Hydrolysed FB_1 (bound)	Kim et al. (2003), Park et al. (2004), Dall'Asta et al. (2008)
	Rice cultures		Fumonisin-fatty acid esters	Motta and Scott (2009)
2.	Maize cultures	Trichothecenes (DON)	D3G	Boutigny et al. (2008)
	Cereal commodities		3-acetyl DON (3-DON)	Berthiller et al. (2005)
	Wheat flour and semolina, oat flour, barley beer, barley malt		15-acetyl DON (15 DON)	Lancova et al. (2008)
	Wheat bread, biscuits, pasta		D3G	Vendl et al. (2010)
			D3G	
			D3G (conjugate)	
3.	Oat, wheat	T-2 and HT-2 toxins	T-2 toxin di glucosides and HT-2 toxin di glucosides	McCormick et al. (2015)
4.	Wheat and maize (suspension cultures)	OTA	OTA-α 4R-4-hydroxy OTA	Ruhland et al. (1996)
			4S-4-hydroxyOTA	
5	Cereal-based products and raw cereals	ZEN	α-ZEL and β-ZEL	De Boevre et al. (2012)
	Wheat bread (whole meal), wheat flour, biscuits, flakes, muesli, snack bars		Z14G (conjugate)	Paris et al. (2014)
	Maize meal, bran flakes		Zearalenone-14-sulphate (Z14S)	
	Barley		Z14S, Z14G, β-ZELG	
			Zearalenone-14-*O*-β-glucoside and Zearalenone-16-*O*-β-glucoside	
6.	Apple juice	Patulin	Bound-patulin	Baert et al. (2007)

7.4 Metabolism of Mycotoxin in Animals

In animals, the metabolism of mycotoxins involves reduction, hydrolysis, or oxidation (phase I enzymatic transformation) and glucuronidation and glucosidation (phase II conjugation). The products of phase I reactions have higher toxicity than their parent toxins; on the contrary, phase 2 products are less toxic. The metabolism of mycotoxins in animals takes place in the liver and kidneys thereby differing from the plants in elimination and distribution patterns (Berthiller et al. 2015). Major mycotoxins except FBs are metabolized into less toxic derivatives in the rumen of ruminants, therefore can be present in the milk and other animal-derived products and by-products (Table 7.2).

7.5 Toxicological Profile

The modified metabolites of parent toxins formed by phase I or phase II reactions can be converted into their parent compounds by the action of digestive enzymes and increase the health-related risks. Masked mycotoxins together with parent compounds may cause systemic as well as localized toxicities. Masked mycotoxins are less toxic in comparison to the native mycotoxins; however, some of them have higher absorption and bioavailability than their parent forms and may impose serious health problems (Rychlik et al. 2014). The metabolites formed by the enzymatic and microbial actions get absorbed in the blood of animals. Later on, the residues of these

Table 7.2 Summarizes the modified forms of mycotoxins found in animal-derived products

S. No.	Food product	Native mycotoxin	Metabolite/s	Reference/s
1.	Edible muscles	AFB_1 in rainbow trout (dietary exposure)	AFM_1 and AFL	Nomura et al. (2011)
2.	Muscles and tissues	AFB_1 in *Clarias gariepinus*	AFL and AFB_1 residue	Suzy et al. (2017)
3.	Milk	AFB_1 in goat feed	AFM_1 and AFQ_1	Vidal et al. (2018)
4.	Milk and dairy products	AFB_1 in contaminated feed of ruminants	AFM_1 and aflatoxicol	Peles et al. (2019), Martinez et al. (2019)
5.	Breast milk	AFB_1 in human	AFM_1 and AFQ_1	Vidal et al. (2018)
6.	Meat and meat by products	OTA in animals	Hydroxylated OTA and OTB	Zhang et al. (2019)
7.	Milk, edible tissues, liver	FBs in bovine	Hydrophilic metabolite; FB_1	Mahmoud et al. (2019), Fink-Fink-Gremmels and Van der Merwe (2019)
8.	Animal tissues and by products	ZEN in contaminated animal feed	α-ZAL (-α-zearalenol) and β-ZAL (β-zearalanol)	Dänicke and Winkler (2015)
9.	Muscles	ZEN in the feed of common carp	α-ZEL	Pietsch et al. (2014)
10.	Eggs	ENN in broiler feed	ENN B	Fraeyman et al. (2016)

metabolites are transferred to the liver as plasma proteins (Vila-Donat et al. 2020). Likewise, one of the metabolites of ZEN, ZEN-4-Glu can hydrolyse in the stomach of pigs (Gareis et al. 1990). Destruxins, the metabolites from the entomopathogenic fungus *Metarrhizium anisopliae* have been found toxic to plants and some vertebrates. The cytotoxicity of destruxins has also been reported on vertebrate cell lines (Pedras et al. 2002). The neurotoxic effect of fusaric acid on rats was reported by Porter et al. (1995). The additive effect of Z14G with ZEN causes potential health hazards in swine and humans (JECFA, WHO, Geneva 2000).

7.6 Detection Methods

The occurrence of masked mycotoxins in food and feed can be detected using three categories of analytical methods:

1. LC-MS/MS-based methods—These methods are quantitative methods and can quantitate the masked mycotoxins along with their parent compounds (Berthiller et al. 2015).
2. Immunochemistry—Mycotoxins can detect based on the cross-reactivity of mycotoxin-specific antibodies (like ELISA).
3. Hydrolysis—The third category is based on the hydrolysis of masked mycotoxins to the parent compound. The hydrolysis reactions can be alkaline, acidic, or enzymatic (Berthiller et al. 2015).

7.7 Future Prospect and Conclusion

Masked mycotoxins are hidden danger thus their occurrence and associated risk could not be undermined. There is a dearth of data on the masked mycotoxins and their toxicological profiles. Therefore, more studies should be done for a better understanding of the threats imposed by them. Furthermore, the risk management strategies like regular testing of food, feed and their products, permissible limits of exposure, and regulatory framework should be implemented for masked mycotoxins also. The application of highly sensitive methods such as biosensors and biomarkers should be extended for the detection of modified mycotoxins as well.

References

Baert K, De Meulenaer B, Kasase C, Huyghebaert A, Ooghe W, Devlieghere F (2007) Free and bound patulin in cloudy apple juice. Food Chem 100(3):1278–1282

Berthiller F, Dall'Asta C, Schuhmacher R, Lemmens M, Adam G, Krska R (2005) Masked mycotoxins: determination of a deoxynivalenol glucoside in artificially and naturally contaminated wheat by liquid chromatography-tandem mass spectrometry. J Agric Food Chem 53:3421–3425

Berthiller F, Crews C, Dall'Asta C, Saeger SD, Haesaert G, Karlovsky P, Oswald IP, Seefelder W, Speijers G, Stroka J (2013) Masked mycotoxins: a review. Mol Nutr Food Res 57:165–186

Berthiller F, Maragos CM, Dall'Asta C (2015) Introduction to masked mycotoxins. In: Masked mycotoxins in food: formation, occurrence and toxicological relevance. RSC, London, pp 1–13. https://doi.org/10.1039/9781782622574-00001. eISBN: 978-1-78262-257-4

Boutigny A-L, Richard-Forget F, Barreau C (2008) Natural mechanisms for cereal resistance to the accumulation of *Fusarium* trichothecenes. Eur J Plant Pathol 121:411–423. https://doi.org/10.1007/s10658-007-9266-x

Coleman J, Blake-Kalff M, Davies E (1997) Detoxification of xenobiotics by plants, chemical modification and vacuolar compartmentation. Trends Plant Sci 2:144–151

Cummins I, Dixon DP, Freitag-Pohl S, Skipsey M, Edwards R (2011) Multiple roles for plant glutathione transferases in xenobiotic detoxification. Drug Metab Rev 43(2):266–280. https://doi.org/10.3109/03602532.2011.552910

Dall'Asta C, Galaverna G, Aureli G, Dossena A, Marchelli R (2008) A LC/MS/MS method for the simultaneous quantification of free and masked fumonisins in maize and maize-based products. World Mycotoxin J 1(3):237–246

Dänicke S, Winkler J (2015) Invited review: diagnosis of zearalenone (ZEN) exposure of farm animals and transfer of its residues into edible tissues (carry over). Food Chem Toxicol 84:225–249. https://doi.org/10.1016/j.fct.2015.08.009

De Boevre M, Di Mavungu JD, Landschoot S, Audenaert K, Eeckhout M, Maene P, Haesaert G, De Saeger S (2012) Natural occurrence of mycotoxins and their masked forms in food and feed products. World Mycotoxin J 5:207–219

Diary Global (n.d.). https://www.dairyglobal.net/health-and-nutrition/nutrition/emerging-masked-mycotoxins-pose-threats-to-livestock/

European Commission (2007) Commission Regulation (EC) No. 1126/2007 on maximum levels for certain contaminants in foodstuffs as regards Fusarium toxins in maize and maize products. Off J Eur Union 255:14–17

Falavigna C, Cirlini M, Galaverna G, Dall'Asta C (2012) Masked fumonisins in processed food: co-occurrence of hidden and bound forms and their stability under digestive conditions. World Mycotoxin J 5:325–334. https://doi.org/10.3920/WMJ2012.1403

Fink-Gremmels J, Van der Merwe D (2019) Mycotoxins in the food chain: contamination of foods of animal origin. In: Fink-Gremmels J, Van der Merwe D (eds) Chemical hazards in foods of animal origin. Wageningen Academic Publishers, Wageningen, pp 1190–1198. https://doi.org/10.3920/978-90-8686-877-3

Fraeyman S, Devreese M, Antonissen G, De Baere S, Rychlik M, Croubels S (2016) Comparative oral bioavailability, toxicokinetics, and biotransformation of enniatin B1 and enniatin B in broiler chickens. J Agric Food Chem 64(38):7259–7264. https://doi.org/10.1021/acs.jafc.6b02913

Galaverna G, Dall'Asta C, Mangi M, Dossena A, Marchelli R (2009) Masked mycotoxins: an emerging issue for food safety. Czech J Food Sci 27:S89–S92. https://doi.org/10.17221/1064-CJFS

Gareis M, Bauer J, Thiem J, Plank G, Grabley S, Gedek B (1990) Cleavage of zearalenone-glycoside, a "masked" mycotoxin, during digestion in swine. J Veterinary Med Ser B 37(3):236–240. https://doi.org/10.1111/j.1439-0450

JECFA in: Joint FAO/WHO Expert Committeee on Food Additives (2000) Safety evaluation of certain food additives and contaminants, WHO/FAO Food Additives Series 44. WHO, Geneva

Kim EK, Scott PM, Lau BP (2003) Hidden fumonisin in corn flakes. Food Addit Contam 20(2):161–169. https://doi.org/10.1080/0265203021000035362. Erratum in: Food Additives & Contaminants. 2003;20(4):417

Lancova K, Hajslova J, Poustka J, Krplova A, Zachariasova M, Dostalek P, Sachambula L (2008) Transfer of *Fusarium* mycotoxins and 'masked' deoxynivalenol (deoxynivalenol-3-glucoside)

from field barley through malt to beer. Food Addit Contam A Chem Anal Cont Expos Risk Assess 25(6):732–744. https://doi.org/10.1080/02652030701779625

Mahmoud MA, Ghattas TA, Amal ZAL, Abed El AMZ (2019) Effect of mycotoxin on reproductive performance in dairy cattle. https://en.engormix.com/mycotoxins/articles/effect-mycotoxin-reproductive-performance-t43882.htm

Martinez MP, Magnoli AP, Pereyra MG, Cavaglieri L (2019) Probiotic bacteria and yeasts adsorb aflatoxin M_1 in milk and degrade it to less toxic AFM_1-metabolites. Toxicon 172:1–7. https://doi.org/10.1016/j.toxicon.2019.10.001

Masri MS, Lundin RE, Page JR, Garcia VC (1967) Crystalline aflatoxin M_1 from urine and milk. Nature 215:753

McCormick SP, Kato T, Maragos CM, Busman M, Lattanzio VM, Galaverna G, Dall'Asta C, Crich D, Price NP, Kurtzman CP (2015) Anomericity of T-2 toxin-glucoside: masked mycotoxin in cereal crops. J Agric Food Chem 63(2):731–738. https://doi.org/10.1021/jf504737f

Motta EL, Scott PM (2009) Bioaccessibility of total bound fumonisin from corn flakes. Mycotox Res 25(4):229–232. https://doi.org/10.1007/s12550-009-0032-0

Nomura H, Ogiso M, Yamashita M, Takaku H, Kimura A, Chikasou M (2011) Uptake by dietary exposure and elimination of aflatoxins in muscle and liver of rainbow trout (*Oncorhynchus mykiss*). J Agric Food Chem 59:5150–5158. https://doi.org/10.1021/jf1047354

Paris MPK, Schweiger W, Hametner C, Stückler R, Muehlbauer GJ, Varga E, Krska R, Berthiller F, Adam G (2014) Zearalenone-16-O-glucoside: a new masked mycotoxin. J Agric Food Chem 62(5):1181–1189. https://doi.org/10.1021/jf405627d

Park JW, Scott PM, Lau BP, Lewis DA (2004) Analysis of heat-processed corn foods for fumonisins and bound fumonisins. Food Addit Contam 21(12):1168–1178. https://doi.org/10.1080/02652030400021873

Pedras MS, Irina Zaharia L, Ward DE (2002) The destruxins: synthesis, biosynthesis, biotransformation, and biological activity. Phytochemistry 59(6):579–596. https://doi.org/10.1016/s0031-9422(02)00016-x

Peles F, Sipos P, Győri Z, Pfliegler WP, Giacometti F, Serraino A, Pagliuca G, Gazzotti T, Pócsi I (2019) Adverse effects, transformation and channeling of aflatoxins into food raw materials in livestock. Front Microbiol 10:2861. https://doi.org/10.3389/fmicb.2019.02861

Pietsch C, Schulz C, Rovira P, Kloas W, Burkhardt-Holm P (2014) Organ damage and hepatic lipid accumulation in carp (*Cyprinus carpio* L.) after feed-borne exposure to the mycotoxin, deoxynivalenol (DON). Toxins (Basel) 6(2):756–778. https://doi.org/10.3390/toxins6020756

Porter JK, Bacon CW, Wray EM, Hagler WM Jr (1995) Fusaric acid in *Fusarium moniliforme* cultures, corn, and feeds toxic to livestock and the neurochemical effects in the brain and pineal gland of rats. Nat Toxins 3(2):91–100. https://doi.org/10.1002/nt.2620030206

Ruhland M, Engelhardt G, Wallnoefer PR (1996) Transformation of the mycotoxin ochratoxin A in plants. 2. Time course and rates of degradation and metabolite production in cell suspension cultures of different crop plants. Mycopathologia 134:97–102

Rychlik M, Humpf H-U, Marko D, Dänicke S, Mally A, Berthiller F, Klaffke H, Lorenz N (2014) Proposal of a comprehensive definition of modified and other forms of mycotoxins including "masked" mycotoxins. Mycotox Res 30:197–205. https://doi.org/10.1007/s12550-014-0203-5

Sewald N, Lepschy von Gleissenthall J, Schuster M, Müller G, Aplin RT (1992) Structure elucidation of a plant metabolite of 4-desoxynivalenol. Tetrahedron 3:953–960

Suzy SK, Thomas EE, Raphael KJ, Christelle TT, Peguy TA, Joseph T (2017) Effect of aflatoxin B1 on growth performance of *Clarias gariepinus* fry (Burchell, 1822) in west Cameroon. Int J Agron Agric Res 10(4):33–41

Vendl O, Crews C, MacDonald S, Krska R, Berthiller F (2010) Occurrence of free and conjugated *Fusarium* mycotoxins in cereal-based food. Food Addit Contam A Chem Anal Cont Expos Risk Assess 27(8):1148–1152. https://doi.org/10.1080/19440041003801166

Veprikova Z, Vaclavikova M, Lacina O, Dzuman Z, Zachariasova M, Hajslova J (2012) Occurrence of mono- and di-glycosylated conjugates of T-2 and HT-2 toxins in naturally contaminated cereals. World Mycotoxin J 5(3):231–240. https://doi.org/10.3920/WMJ2012.1453

Vidal A, Mengelers M, Yang S, De Saeger S, De Boevre M (2018) Mycotoxin biomarkers of exposure: a comprehensive review. Compr Rev Food Sci Food Saf 17:1127–1155. https://doi.org/10.1111/1541-4337.12367

Vila-Donat P, Marín S, Sanchis V, Ramos AJ (2020) Tri-octahedral bentonites as potential technological feed additive for *Fusarium* mycotoxin reduction. Food Addit Contam A Chem Anal Cont Expos Risk Assess 37(8):1374–1387. https://doi.org/10.1080/19440049.2020.1766702

Zhang Z, Fan Z, Nie D, Zhao Z, Han Z (2019) Analysis of the carry-over of ochratoxin A from feed to milk, blood, urine, and different tissues of dairy cows based on the establishment of a reliable LC-MS/MS method. Molecules 24:2823. https://doi.org/10.3390/molecules24152823

Chapter 8
Extraction, Characterization, and Analysis of Mycotoxins

8.1 Introduction

Mycotoxins are secondary metabolites of fungus that have a low molecular weight and pose a threat as toxic contaminants of food products, thus making its effective monitoring and management essential. Around 25% of the world's food crops are contaminated with mycotoxins. Mycotoxin contamination causes serious implications on the quality and safety of food and feed with adverse affects on international trade. Maximum concentration of mycotoxins is present in grain and food based on grains. Mycotoxin contamination depends on several factors such as moisture content in air, temperature, season of harvest, pre- and post-harvest methods of handling, storage and transportation. Advanced methods of mycotoxin detection and analyses show that almost 100% of the crops are contaminated with one or more mycotoxins (De Saeger et al. 2016; Stadler et al. 2020).

In the future, industries might witness new challenges of mycotoxins as mycotoxins often occur as masked or bound mycotoxins. Mycotoxins are stable under room temperature and neutral conditions; however, mycotoxins have been known to produce new forms under the action of pressure, temperature, pH, enzymes, and presence of other molecules present in their vicinity. They also form conjugates with other macromolecules such as sugars, proteins, and lipids (De Saeger et al. 2016; Zhang et al. 2018).

In order to fulfil the legislative requirements and for the protection of consumer health, developments of detection techniques that are quick screening methods are the need of the hour that can be used outside the laboratory and have a high sensitivity-precision.

K. Singh, A. Kumari, *Mycotoxins and Mycotoxicoses*,
https://doi.org/10.1007/978-981-19-2370-8_8

8.2 Methods of Detection

Sensitive and reliable methods for the detection of mycotoxins are required since the toxicity of mycotoxins occurs at very low concentrations. Figure 8.1 shows the different steps of mycotoxin analysis from sampling to its quantification. To obtain reliable analytical results extraction and preparation of samples are key points since mycotoxins are not homogeneously distributed in commodities. Process of analyses includes extraction, purification, and determination followed by chromatography. Apart from this, fluorodensitometry is also used for mycotoxin determination.

8.2.1 *Extraction*

The foremost step in mycotoxin analysis is extraction. The objective of extraction is to extract maximum possible quantity of mycotoxin from the substrate with the help of a suitable solvent. The extraction procedure should be such that maximum quantity of the mycotoxin is present in the solvent and removed from the substrate

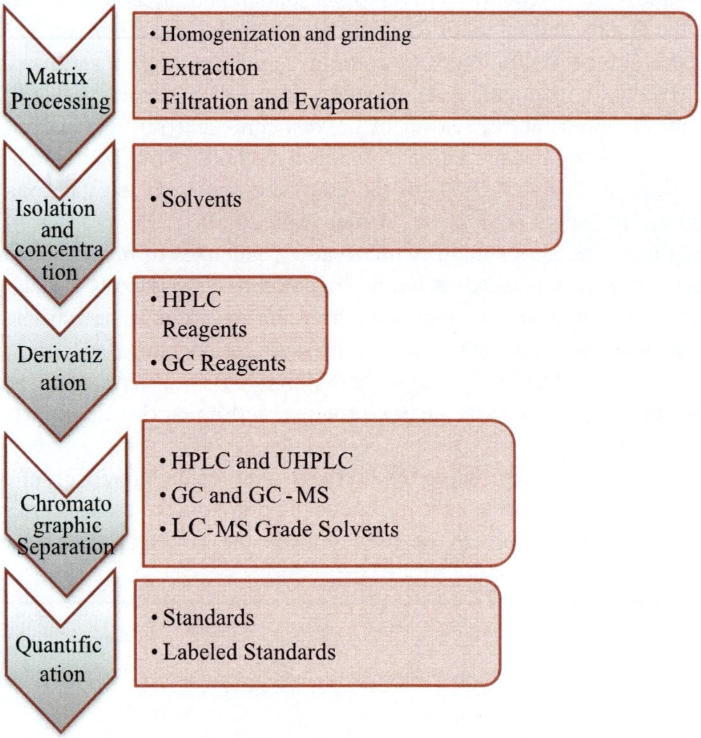

Fig. 8.1 Steps in the analyses of mycotoxins

or the solid phase of the mixture. The recovery of mycotoxin from the mixture depends on the process of extraction (CAST 2003). Recovery rate of mycotoxin is around 70–110% during extraction but the consistency of the process is the main factor. Extraction of mycotoxins from food material is carried out using a mixture of water and comparatively polar solvents like acetonitrile, acetone, ethyl acetate, and methanol. The most widely used solvents are acetonitrile and methanol for extraction. However, for patulin ethyl acetate is chosen as a solvent. The physical and chemical properties of the mycotoxin to be extracted, its substrate, safety considerations, and the analytical procedures are the factors that determine the composition of the solvent used for extraction (Stroka et al. 1999). The composition of the solvent should be such that it is able to extract the maximum possible toxins from foods and is also suitable for the analytical system. For instance, chlorinated solvents, like methylene chloride extract aflatoxins from corn very efficiently but it is insoluble in water thus barring its use in water-based detection tests such as enzyme-linked immunosorbent assays (ELISAs). However, when solid-phase extraction columns are used for aflatoxins extraction, solubility as factor can be excluded. Costs for waste solvent disposal and safety considerations are other important factors that should be considered during solvent selection. Extraction can be performed by either blending at a high speed for a few minutes or shaking with the solvent for half an hour to 2 h. Blending is a quicker process than shaking when the number of samples to be analysed is few. It should be kept in mind that the entire sample should be washed continuously with the extraction solvent. But when the samples are large in number, shaking is preferred to blending as modern shakers are able to hold 20 samples simultaneously. As the shaking or blending of the sample with extraction solvent is over, the liquid phase is separated from the solid either by centrifugation or filtration. Further, cleaning up of the extract takes place or the extracted mixture is directly analysed for the presence of toxins (Pittet 2005).

8.2.2 Chromatographic Techniques

Analytical methods for the detection of mycotoxins are thin-layer chromatography (TLC), gas chromatography (GC), gas chromatography-mass spectrometry (GC-MS), liquid chromatography (LC), liquid chromatography-mass spectrometry (LC-MS), and enzyme-linked immunosorbent assay (ELISA) (Man et al. 2017). TLC being a cheap, quick, and simple technique is, therefore, the most widely used laboratory method for qualitative and quantitative determination. Bidimensional TLC is an option for complex extracts. The use of volatilization in gas-chromatography makes its application limited in mycotoxin analysis. Gas-chromatography coupled with mass spectrometry (GC-MS) can detect as low as 0.0001 ppm of the mycotoxin. (Rahmani et al. 2009, https://www.perkinelmer.com/uk/category/mycotoxins-in-food).

8.2.2.1 High-Performance Liquid Chromatography

Chromatography is the most frequently used method for mycotoxin analysis as it provides higher accuracy and repeatability. HPLC (high-performance liquid chromatography) with ultraviolet (UV) or fluorescence detection (FLD) are the most widely used methods (De Saeger et al. 2016). Analytical methods should be good enough to determine even low concentrations of mycotoxins at least down to regulatory limits.

UV fluorescence detection in combination with HPLC is a widely accepted and used chromatographic technique and is often employed for the analysis of aflatoxins. It has the capacity to identify mycotoxin with a limit of detection (LOD) even below ng/g product. Mycotoxin (https://www.perkinelmer.com/uk/category/mycotoxins-in-food).

8.2.2.2 Reversed-Phase HPLC

Reversed-phase HPLC (RP-HPLC) in combination with fluorescence detection was used for the detection of citrinin in fermented products of *Monascus* using acidified water and acetonitrile as mobile phase. Independent LC-MS/MS analysis methods were developed to identify citrinin in xuezhikang capsules and other 11 fermented *Monascus* products (Xue-Mei et al. 2012).

8.2.2.3 Multichannel UV Detector with HPLC

A new method was developed and confirmed by Curticapean et al. (2011) to simultaneously determine the number of related mycotoxins from seeds of corn. The method combined the use of a multichannel UV detector with HPLC (Dionex UltiMate3000 modular system) for the concurrent detection of aflatoxin B1, zearalenone, and ochratoxin A. Internal standards were primarily used for the development stages. Ethyl acetate was found to be a good organic solvent for the extraction procedure of mycotoxins from the matrix as it results in 60–70% recovery of mycotoxins (Curticapean et al. 2011).

8.2.3 Mass Spectrometry

This method makes use of ambient ionization and allows non-destructive sampling and detection. One such technique for fumonisin B1 analysis in maize is desorption electrospray ionization (DESI) (Maragos and Busman 2010).

Other spectroscopic methods are also implemented in mycotoxin detection like the near-infrared (NIR) spectroscopy and electronic noses. Electronic noses make use of a group of non-specific chemical detectors that combine with volatile molecules to generate signals that are used as markers of the volatile compounds in the sample (De Saeger et al. 2016).

8.2.4 *FTIR and ^{1}H NMR*

In 2010, Singh et al. made use of submerged fermentation of *Alternaria alternata* cultures to isolate tenuazonic acid. Cell-free culture filtrate (CFCF) was used in the isolation and synthesis of tenuazonic acid by *A. alternata*. Solvent partition, thin layer chromatography (TLC), FTIR, and ^{1}H NMR analysis were employed in the purification and characterization of the mycotoxin (Sanodiya et al. 2010).

8.2.5 *New Methods*

Mycotoxins can be directly detected through physical characteristics and indirectly through immunoassays (Maragos and Busman 2010). Some new techniques for rapid identification of mycotoxins have also been described in the literature. They include:

8.2.5.1 Immunological Techniques

Immunoaffinity column-based analyses (IAC) or enzyme-linked immunosorbent assay (ELISA) are two methods performed for monitoring mycotoxins. Immunoassays are the most sensitive, quick, and precise analytical tools for examining mycotoxins in food as well as feed. The bases for immunoassays are the production of monoclonal and polyclonal antibodies that bind to antigenic toxins. For rapid analysis of aflatoxins, fumonisins, ocratoxin, zearelenone, etc., a number of commercial tests are also available (Aldao et al. 1995; Pittet 2005).

8.2.5.2 Biosensors

Biosensors are another rapid mycotoxin screening methods that make use of antibodies adhered to the surface of a sensor in excess. The mycotoxin-conjugate gets immobilized on the surface of the sensor and the alterations in the refractive index are recorded. Other optical principles (evanescent wave fibre optic) are used for the screening of aflatoxins and fumonisins (Strachan et al. 1997; Mascini 2001; Mascini et al. 2001).

8.2.5.3 Capillary Electrophoresis

Capillary electrophoresis along with fluorescence detection is another technique for detecting trace quantities of mycotoxins using laser-induced fluorescence (Maragos et al. 1996).

8.2.5.4 Chemiluminescence and Bioluminescence Assays

Chemiluminescence and bioluminescence assays (Sarter and Zakhia 2004) are other reliable, quick, and sensitive techniques for mycotoxins detection that make use of luminescence. Antigen-antibody binding being detected at the last step of an immuno-enzymatic assay forms the basis of the chemiluminescent method. In the bioluminescence technique, fungi are detected in a test sample on the basis of the reaction between microbial ATP and luciferase to produce luminescence (Oancea and Stoia 2008).

8.2.5.5 Microplate ELISA

Microplate ELISA is another method of mycotoxin characterization, however, being time-consuming as it might take up to several hours. Detection and quantification of aflatoxin B1, zearalenone, and ochratoxin A were performed using this method. This technique combines accuracy and sensitivity providing a rapid method for mycotoxin analysis. The method utilizes biotin along with streptavidin-polyperoxidase conjugate that emits strong signals even under non-equilibrium conditions. Time taken for the assay of one mycotoxin was 20 min while that for three mycotoxins simultaneously was 25 min. Mycotoxin recovered was at least 95% in a water-organic solvent. Successful implementation of the method was performed while testing samples of corn and in poultry products making use of standard mycotoxins as internal standards. Here, the limit of detection (LOD) for aflatoxin B1 was 0.24 ng/g, for ochratoxin A was 1.3 ng/g, and for zearalenone was 3 ng/g (Urusov et al. 2015).

8.2.5.6 Using Mycotoxin Biosynthetic Gene Clustering for Identifying Mycotoxins

Several mycotoxins have been successfully identified using the sequence of gene cluster that synthesizes the mycotoxin. Genes for the synthesis of mycotoxins are present in clusters and some seem to have undergone lateral gene transfer amongst species resulting in its presence in several species (Sidhu 2002). A more common method for detection can be adopted once we come to know the sequence of the genes responsible for the synthesis of a particular mycotoxin. Primers can be developed for the detection of mycotoxins by utilizing the homologous sequences in different mycotoxin biosynthetic genes in different species. As mycotoxin synthesizing genes are present in clusters, detection of one gene can lead to the identification of other genes involved in the pathway (Sadhasivam et al. 2017).

8.2.5.7 Real-Time PCR Assay

A new method has been developed using the *nor-1* gene to detect aflatoxin producing fungi utilizing real-time PCR assay. The method, however, is unable to differentiate between the aflatoxin producing and non-aflatoxin producing isolates. Another limitation of this method is that species of *Aspergillus* producing sterigmatocystin have also not been checked for this method. A gene *aflJ* is known to be involved in the synthesis of aflatoxin but is absent in the sterigmatocystin biosynthesis pathway. This gene can prove to be a good target to differentiate between aflatoxin and sterigmatocystin producing fungi. Two other genes, *pks-A* and *aflR*, have been studied utilizing a real-time RT-PCR assay to check the effect of nutritional factors and inhibitors on the production of aflatoxin (Sadhasivam et al. 2017).

8.2.5.8 Nucleic Acid Sequence-Based Amplification (NASBA)

Nucleic acid sequence-based amplification (NASBA) or self-sustained sequence replication (SSR) is a method that makes use of three enzymes namely AMV RT, Rnase H, and T7 RNA polymerase that amplify specific RNA sequences in an isothermal reaction. Biosynthesis of *Fusarium* mycotoxin, fumonisin can be determined with the help of this technique by identifying the RNA of the *Fum5* gene and trichothecene can be detected by identifying the *Tri5* gene (Nicholson et al. 1998). The only drawback of using this technique is that like RT-PCR, NASBA can also detect the toxin only when the fungi are actively synthesizing the mycotoxin and not when the fungi are growing or are in dormancy when conditions are not favourable for mycotoxin production (Mostafa et al. 2012).

8.2.5.9 Lateral Flow Devices (LFD)

Lateral flow devices (LFD) are quick and accessible devices that are recently being used. They selectively and specifically use antibodies to detect the target mycotoxin and quantify it in a few minutes without the use of any other chemical or costly instruments. LFD is now being widely used in place of ELISA (Wang et al. 2022).

Fluorescence polarization immunoassay (FPIA) and surface plasmon resonance sensors (SPRS) are some other approaches, which also make use of immunochemical methods for the identification of mycotoxins (Li et al. 2021; De Saeger et al. 2016; Wang et al. 2022).

8.3 Conclusion

Since mycotoxins at low concentrations can cause toxicity, therefore, it is necessary that the methods of detection are sensitive and reliable. Another point to be kept in mind is the method of sampling. Sampling and careful analysis should be carried out so that consignments with high mycotoxin concentrations are not accepted and those with levels below the permissible limit are not unnecessarily rejected. Physical and chemical analyses in laboratories are still important methods for the primary analysis. Various other techniques making use of TLC, HPLC, GC, and LC–MS have been employed in mycotoxin detection and analyses. Cleanup methods such as multifunctional columns and immunoaffinity columns have also been established as they avoid the usual column chromatography cleanup or liquid–liquid partitioning. Test kits based on immuno-analytical principles which are rapid and easy-to-use are paving its path for quick detection of mycotoxins in imported and exported products. As more and more sensitive and specific methods are becoming prevalent, inspection of mycotoxin contaminated food is becoming rampant. To validate a method for detection of mycotoxins, it should be reliable, specific, and rapid.

References

Aldao MA, Carpinella MC, Corelli M, Herrero GG (1995) Competitive ELISA for quantifying small amounts of aflatoxin B1. Food Agric Immunol 7(4):307–314. https://doi.org/10.1080/09540109509354890

Council for Agricultural Science and Technology (2003) Mycotoxins: risks in plant, animal, and human systems. Council for Agricultural and Technology, Ames

Curticapean A, Toma F, Tarcea M, Curticapean M, Samarghitan V, Pop IA, Gulea A (2011) HPLC method validation for simultaneous determination of three mycotoxins from corn seeds. Croat Chem Acta 84(3):413–417

De Saeger S, Audenaert K, Croubels S (2016) Report from the 5th international symposium on mycotoxins and toxigenic moulds: challenges and perspectives (Mytox) held in Ghent, Belgium. https://doi.org/10.3390/toxins8050146

Li R, Wen Y, Wang F, He P (2021) Recent advances in immunoassays and biosensors for mycotoxins detection in feedstuffs and foods. J Anim Sci Biotechnol 12(1):1–9

Man Y, Liang G, Li A, Pan L (2017) Analytical methods for the determination of Alternaria mycotoxins. Chromatographia 80(1):9–22. https://doi.org/10.1007/s10337-016-3186-x

Maragos CM, Bennett GA, Richard JL (1996) Analysis of fumonisin B 1 in corn by capillary electrophoresis. In: Fumonisins in food. Springer, Boston, pp 105–112. https://doi.org/10.1007/978-1-4899-1379-1_9

Maragos CM, Busman M (2010) Rapid and advanced tools for mycotoxin analysis: a review. Food Addit Contam 27(5):688–700. https://doi.org/10.1080/19440040903515934

Mascini M, Palchetti I, Marrazza G (2001) DNA electrochemical biosensors. Fresenius J Anal Chem 369(1):15–22

Mascini M (2001) Affinity electrochemical biosensors for pollution control. Pure Appl Chem 73(1): 23–30. https://doi.org/10.1351/pac200173010023

Mostafa AT, Armin A, Hamid P, Reza AM (2012) Rapid detection methods for analysis of fungi and mycotoxins in agriculture products. Res J Recent Sci 1(7):90–98

Nicholson P, Simpson DR, Weston G, Rezanoor HN, Lees AK, Parry DW, Joyce D (1998) Detection and quantification of *Fusarium culmorum* and *Fusarium graminearumin* cereals using PCR assays. Physiol Mol Plant Pathol 53(1):17–37. https://doi.org/10.1006/pmpp.1998.0170

Oancea S, Stoia M (2008) Mycotoxins: a review of toxicology, analytical methods and health risks. Univ Cibiensis Ser E 7(1):19–36

PerkinElmer. Discover food safety and quality. https://www.perkinelmer.com/uk/category/mycotoxins-in-food. Accessed 12 Dec 2021

Pittet A (2005) Modern methods and trends in mycotoxin analysis. Mitt Lebensmittelunters Hyg 96(6):424–444

Rahmani A, Jinap S, Soleimany F (2009) Qualitative and quantitative analysis of mycotoxins. Compr Rev Food Sci Food Saf 8(3):202–251. https://doi.org/10.1111/j.1541-4337.2009.00079.x

Sadhasivam S, Britzi M, Zakin V, Kostyukovsky M, Trostanetsky A, Quinn E, Sionov E (2017) Rapid detection and identification of mycotoxigenic fungi and mycotoxins in stored wheat grain. Toxins 9(10):302. https://doi.org/10.3390/toxins9100302

Sanodiya BS, Thakur GS, Baghel RK, Pandey AK, Prasad GB, Bisen PS (2010) Isolation and characterization of tenuazonic acid produced by *Alternaria alternata*, a potential bioherbicidal agent for control of *Lantana camara*. J Plant Prot Res 50(2):133–139. https://doi.org/10.2478/v10045-010-0023-3

Sarter S, Zakhia N (2004) Chemiluminescent and bioluminescent assays as innovative prospects for mycotoxin determination in food and feed. Lumin J Biol Chem Lumin 19(6):345–351. https://doi.org/10.1002/bio.791

Sidhu GS (2002) Mycotoxin genetics and gene clusters. Eur J Plant Pathol 108(7):705–711

Stadler D, Berthiller F, Suman M, Schuhmacher R, Krska R (2020) Novel analytical methods to study the fate of mycotoxins during thermal food processing. Anal Bioanal Chem 412(1):9–16

Strachan NJ, John PG, Millar IG (1997) Application of an automated particle based immunosensor for the detection of aflatoxin B1 in foods. Food Agric Immunol 9(3):177–183. https://doi.org/10.1080/09540109709354948

Stroka J, Petz M, Joerissen U, Anklam E (1999) Investigation of various extractants for the analysis of aflatoxin B1 in different food and feed matrices. Food Addit Contam 16(8):331–338. https://doi.org/10.1080/026520399283902

Urusov AE, Zherdev AV, Petrakova AV, Sadykhov EG, Koroleva OV, Dzantiev BB (2015) Rapid multiple immunoenzyme assay of mycotoxins. Toxins 7(2):238–254. https://doi.org/10.3390/toxins7020238

Wang Y, Zhang C, Wang J, Knopp D (2022) Recent progress in rapid determination of mycotoxins based on emerging biorecognition molecules: a review. Toxins 14(2):73. https://doi.org/10.3390/toxins14020073

Xue-Mei L, Xing-Hai S, Lan X, Zhen-Wen D, Shu-Ren G (2012) A validated RP-HPLC method for the determination of citrinin in xuezhikang capsule and other Monascus-fermented products. E-J Chem 9(1):260–266. https://doi.org/10.1155/2012/693570

Zhang L, Dou XW, Zhang C, Logrieco AF, Yang MH (2018) A review of current methods for analysis of mycotoxins in herbal medicines. Toxins 10(2):65. https://doi.org/10.3390/toxins10020065

Chapter 9
Management Strategies

9.1 Introduction

Mycotoxin contamination in food and animal feed is a cause of concern globally. The mycotoxins present in the indoor air also pose serious health problems in humans. These hidden threats are heat stable and unavoidable. Mycotoxins can tolerate physical, chemical, and biological methods of food processing. They contribute significantly to the virulence of their producers, the mycotoxigenic fungi that can contaminate crops in pre-harvest, harvest, post-harvest, and stored conditions (FAO 2004). About 300 mycotoxins are recognized so far, out of which only a few exert toxic effects on livestock and human health in the form of primary mycotoxicosis (WHO 2018). The entry of mycotoxins into the food chain and their bioaccumulation leads to long-term effects (secondary mycotoxicosis). The non-specific symptoms of mycotoxicity make its diagnosis difficult. Besides, the interactions of their etiological agents with other stressful conditions like insect infestation worsen the situation (Lopez-Garcia et al. 1999). The emergence of new mycotoxins, the presence of masked mycotoxins, and mycotoxin co-occurrence are the big challenges for the food industry and public health today.

9.2 Mycotoxins in Food and Feed

Healthy food is a need for all the time. However, amid the COVID-19 pandemic, it has become necessary to avoid food-borne illnesses. Many countries set up regulations on toxins produced by algae, fungi, and plants (phycotoxins, mycotoxins, and phytotoxins). Among these natural toxins, mycotoxins are the cause of concern for food safety. Because of this, many developed and a few developing countries impose food safety systems and other regulations. Unlike developed countries, dependency on a single cereal food often leads to dietary exposure of mycotoxin more than TDI

K. Singh, A. Kumari, *Mycotoxins and Mycotoxicoses*,
https://doi.org/10.1007/978-981-19-2370-8_9

(maximum tolerable daily intake) and people are compelled to consume low quality food due to food scarcity in developing countries (Shephard et al. 2019). Moreover, food insufficiency is the major hindrance in implementing food safety programmes and legislation enforcement in these countries. Thus, a clear cut demarcation exists between standards of mycotoxins risk assessment in the developed and developing worlds.

9.3 Factors Responsible for Mycotoxin Production

Some "time-bound factors" control mycotoxin production in pre-harvest, post-harvest, and stored agricultural commodities (Magan et al. 2011; Adegoke and Letuma 2013). They are

1. Extrinsic factors like temperature, climatic conditions and O_2 level, unseasonal rains, high temperatures, and flash floods during harvesting.
2. Intrinsic factors such as type of plant, substrate, water activity, moisture content, and nutrient composition.
3. Implicit factors viz. insect infestation, interaction with other fungal species or other microbes.
4. Processing factors like handling, drying, cleaning, and blending of grains.
5. Preservation using additives.

9.4 Decontamination

Mycotoxins contamination in a wide variety of raw and processed food and feed is inevitable; however, the risk of their contamination can be minimized using different decontamination strategies (Čolović et al. 2019). Of note, these methods should not affect the edibility and nutritive value of the food and should not have toxic by-products or residues (Pankaj et al. 2018). A cost-effective and simple approach is needed for the mitigation of mycotoxins (Huss et al. 2018). Table 9.1 summarizes mycotoxin decontamination methods.

9.5 Prevention

Mycotoxin contamination is a challenge to food safety (Lopez-Garcia et al. 1999). The multigrain composition of food and feed leads to dietary exposure of a mycotoxin cocktail in humans and animals. The long term exposure of multi-mycotoxin (although in the lower cone) is unpredictable. But, owing to their heterogeneity and chemical diversity, the mitigation of mycotoxins is not possible by a single method. Therefore, an integrated approach is needed for mycotoxin management in food and

Table 9.1 Mycotoxin decontamination methods

S. No.	Strategy/process for decontamination	Processing method/s	Reference/s
1.	Cleaning of grains		
	(a) Cleaning	1. Screening to remove under or oversized grain particles	Schofield (2005)
		2. Removal of dust, crushed and fungus infested kernels	Bullerman and Bianchini (2007)
	(b) Milling process	1. Processing of edible parts and inedible by-products separately	Vanara et al. (2018)
		2. Processing of whole grains using roller mill or hammer mill or disk mill or pulverizer	Thomas et al. (2018)
	(c) Separation system (based on screening, gravity and air-separation)	1. Using several classifiers to remove from fine to coarse dust particles	Vidosavljević et al. (2018)
		2. Removal of broken kernels	
	(d) Dry cleaning	1. Polishing grain surface using brushes (having polypropylene bristles)	Vidosavljević et al. (2018)
	(e) Sorting (manual or optic)	1. To separate same sized healthy kernels from the contaminated kernels	Čolović et al. (2019), Karlovsky et al. (2016)
2.	Thermal processing at 150 °C or more	1. Roasting	Kabak (2009)
		2. Extrusion processing	
3.	Feed additives	1. Use of microorganisms, yeast cell wall, mineral clay for detoxification	Kolosova and Stroka (2011)
		2. Reduction in mycotoxin absorbance using absorbents or binders like zeolite, bentonite, phyllosilicates (kaolinites, illite, and smectite clay), tectosilicates, activated charcoal (non-specific sequester), cellulose, peptidoglycans, glucomannans, synthetic polymers (polyvinylpyrrolidone and cholestryramine)	EFSA (2005, 2010, 2022)
		4. Carbon-based nano-particles	Horky et al. (2018)
		5. Sequesters like wheat straw, alfalfa hay, grape pomace	Stangroom and Smith (1984), Greco et al. (2019)
		6. Humic acid derivatives and live bacteria (lactic acid) and yeast (*Saccharomyces cerevisiae*)	Jard et al. (2009)
4.	Dietary supplements and antioxidants	Provide protection against mycotoxins	Sorrenti et al. (2013); Kumari and Singh (2021)

feed. The hazard analysis and critical control point (HACCP) is one such approach that includes prevention and control of mycotoxin production in agricultural commodities from pre-harvest to storage (Lopez-Garcia and Park 1998).

1. Pre-harvest management is the best preventive measure. Fungal contamination in pre-harvest crops plays a crucial role in mycotoxin production and often amplifies in successive stages (harvest, post-harvest, and storage). The pre-harvest system regulates mycotoxin production in field crops by controlling critical factors like insect infestation. It includes crop rotation, proper management of irrigation and crop residues, crop rotation, soil fertility, use of fungus-resistant plant varieties and high temperatures (drought conditions) (https://www.fao.org/3/y1390e/y1390e0c.htm).
2. Timely harvesting reduces the production of mycotoxin in field crops. Besides, proper cleaning and effective drying are significant in mycotoxin mitigation (Lopez-Garcia et al. 1999).
3. Various non-invasive decontamination procedures are included in post-harvest management (Table 9.1).

Irradiation, proper aeration and adequate dryness inhibit fungal growth and mycotoxin formation in stored crops. Appropriate infrastructure for segregation and efficient transport facilities also contribute significantly to mycotoxin-associated risk management.

9.6 Management

Mycotoxin management depends upon mycotoxin-associated risk assessment (FAO 2004). The regulatory agencies like JECFA (the Joint Expert Committee on Food Additives), EFSA (2010) (European Food Safety Authority) set guidelines for the assessment and safety evaluation of contaminants including mycotoxins (FAO, EFSA). Safety measures recommended by JECFA include the identification and characterization of potential hazards. Based on JECFA recommendations, the permissible limit of mycotoxins in food products and feeding stuffs are determined by the Codex Alimentarius Commission (WHO 2018). The data are evaluated in PIWI (Provisional Tolerable Weekly Intake) or PIDI (Provisional Tolerable Daily Intake). However, this approach is not applicable for mycotoxins having genotoxicity such as aflatoxins. In this case, the tolerable intake level should be As Low As Reasonably Achievable (ALARA). The level of mycotoxin contamination can also be measured by its dietary exposure. Besides, the reliable and validated analytical methods, certified reference materials, and proficiency testing schemes like FAPAS (Food Analysis Performance Assessment Scheme) also have significant role in the AQA (Analytical Quality Assurance) criterion (CEC 2006). Further, the incorrect sampling may lead to the wrong evaluation of mycotoxin. Therefore, utmost care should be taken in sample collection.

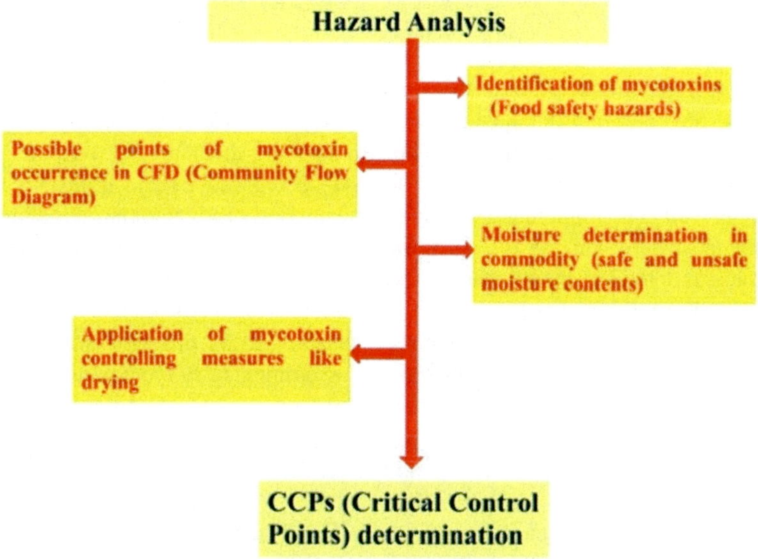

Fig. 9.1 Application of HACCP in mycotoxin hazard analysis. (Source—An overview of HACCP, FAO)

Several countries have implemented the harmonized regulations with their trade contacts for some major and few emerging mycotoxins. The first regulation was implemented for aflatoxin in the 1950s. Since then, several inquiries have been conducted and many recommendations have been published frequently on mycotoxin regulations by many international agencies, including WHO (WHO 2018).

The risk associated with mycotoxin contamination could be managed through HACCP. Based on the engineering system, FMEA (failure mode and effect analysis), this system adopts a holistic approach (identification, evaluation, and control of hazardous substances) for food safety. Its implementation, however, must be preceded by good agricultural practices (GAP) including good hygienic practices (GHP) and good storage practices (GSP) and GMP (good manufacturing practices), such as suitable design of food premises with proper maintenance and effective measures to reduce the risk of mycotoxin contamination, sanitation, personal hygiene of food handlers, means of transportation, adequate information, and awareness about the product. The HACCP team completes the task of mycotoxin hazard analysis and control in five stages (Fig. 9.1).

In silage, mycotoxin contamination can be prevented by good agricultural practices, timely harvest, enhanced fermentation, elimination of oxygen, rapid and tight filling of silos (Whitlow and Hagler 2006).

9.7 Regulation

The role of mycotoxin intoxication in humans and animals has been recognized globally. Several countries have established regulations on mycotoxin contamination to minimize the health risk. Though most of them are focused on major mycotoxins, few countries especially EU set up regulations and limits for minor toxins as well. The implementation and proclamation of mycotoxin regulations can be affected by low or unavailability of their exposure and toxicity data, sensitive methods for detection, quantity and distribution of food and feed and insufficiency of food (van Egmond et al. 2007).

Earlier, regulations on mycotoxin were established by different countries separately. Nowadays, the hazard and exposure assessments of mycotoxins in food are evaluated by JECFA (Joint FAO/WHO Expert Committee on Food Additives). The maximum tolerable limit of mycotoxins in food and feed has been set by an intergovernmental standards-setting body (the Codex Alimentarius Commission) on the recommendations of JECFA (FAO 1997, 2004; WHO 2018). The levels set by the Codex serve as standards for international trade and food supplies (WHO 2018). Figure 9.2 depicts the permissible limits of various mycotoxins in food and feed of the developed and developing countries.

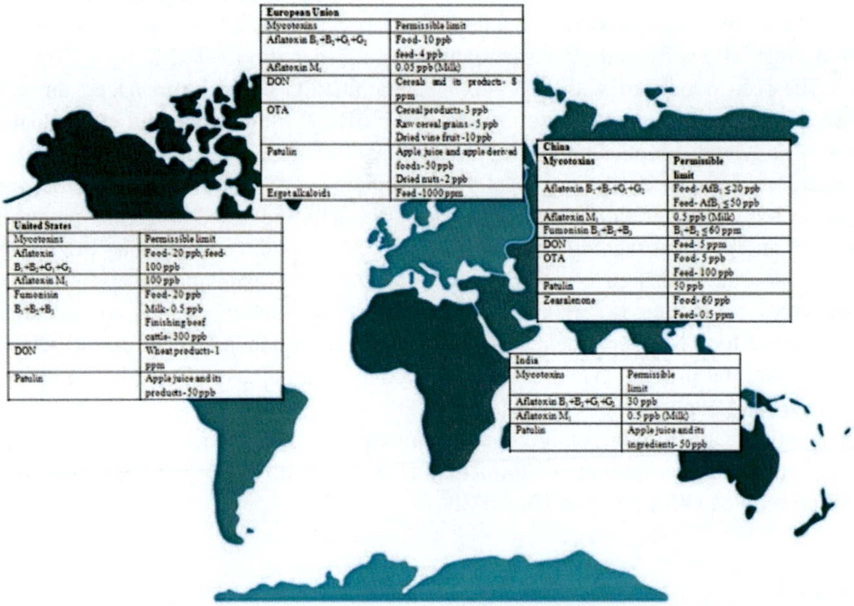

Fig. 9.2 Mycotoxin-permissible limits in food and feed of the developed and developing countries

9.8 Mycotoxin Reduction at Consumer Level: Suggestions

To minimize dietary exposure to mycotoxins, consumers are advised to discard food products having fungal growth or discolouration or insect infestation. It is necessary to consume fresh food. The proper dryness should be maintained in the storage area. Food commodities and nuts should not be kept long. To reduce the risk of mycotoxin exposure, the intake of diverse diets is recommended (WHO 2018). The consumers should be aware of food safety information and risk (Thio and Wijaya 2010).

9.9 Conclusions

As mycotoxin production in agricultural commodities is inexorable, thus decontamination methods and preventive measures (including good agricultural practices, in-field crop protection and proper storage) have an important role to play in mycotoxin mitigation. Once formed, they are hard to remove and remain in the "plough to plate" chain. Mycotoxin occurrence is a global problem. Therefore, the harmonized limits and regulations for mycotoxins have been established by many international regulatory agencies like FAO, EFSA. However, due to limited resources and subsistence farming, food protection is compromised in the developing world. Further, there is a need to address new threats for food safety like climatic change and its impact on mycotoxin production, the emergence of new and modified toxins and unpredicted interactions of mycotoxins.

References

Adegoke GO, Letuma P (2013) Strategies for the prevention and reduction of mycotoxins in developing countries. In: Makun HA (ed) Mycotoxin and food safety in developing countries, London, pp 123–136. https://doi.org/10.5772/52542

Bullerman LB, Bianchini A (2007) Stability of mycotoxins during food processing. Int J Food Microbiol 119:140–146

Commission of the European Communities (CEC) (2006) Commission Regulation (EC) No. 401/2006 of 23 February 2006 laying down the methods of sampling and analysis for the official control of the levels of mycotoxins in foodstuffs. Off J Eur Union L70:12–34

Čolović R, Puvača N, Cheli F, Avantaggiato G, Greco D, Đuragić O, Kos J, Pinotti L (2019) Decontamination of mycotoxin-contaminated feedstuffs and compound feed. Toxins (Basel). 11(11):617. https://doi.org/10.3390/toxins11110617

European Food Safety Authority (EFSA) (2005) Opinion of the scientific panel on additives and products or substances used in animal feed (FEEDAP) on the safety of the product "Biomin BBSH 797" for piglets, pigs for fattening and chickens for fattening. EFSA J 3(2):69

European Food Safety Authority (EFSA) (2010) Scientific opinion on establishing food-based dietary guidelines. EFSA J 8(3):1460

European Food Safety Authority (EFSA) (2022) Statement on Guidelines for 'Substances for Reduction of the Contamination of Feed by Mycotoxins'. https://www.efsa.europa.eu/en/efsajournal/pub/1693. Accessed 21 Jan 2022

Food and Agriculture Organization (1997) Worldwide regulations for mycotoxins 1995. A compendium. FAO food and nutrition paper 64. Food and Agriculture Organization of the United Nations, Rome

Food and Agriculture Organization (2004) Worldwide regulations for mycotoxins in food and feed in 2003. FAO food and nutrition paper 81. Food and Agriculture Organization of the United Nations, Rome

Greco D, D'Ascanio V, Santovito E, Logrieco AF, Avantaggiato G (2019) Comparative efficacy of agricultural by-products in sequestering mycotoxins: multi-mycotoxin adsorption efficacy of agricultural by-products. J Sci Food Agric 99:1623–1634

Horky P, Skalickova S, Baholet D, Skladanka J (2018) Nanoparticles as a solution for eliminating the risk of mycotoxins. Nanomaterials 8:727

Huss A, Cochrane R, Muckey M, Jones C (2018) Animal feed mill biosecurity. In: Food and feed safety systems and analysis. Elsevier, Amsterdam, pp 63–81. ISBN 978-0-12-811835-1

Jard G, Liboz T, Mathieu F, Guyonvarćh A, Lebrihi A (2009) Adsorption of zearalenone by *Aspergillus japonicus* conidia: new trends for biological decontamination in animal feed. World Mycotoxin J 2:391–397

Kabak B (2009) The fate of mycotoxins during thermal food processing. J Sci Food Agric 89:549–554

Karlovsky P, Suman M, Berthiller F, De Meester J, Eisenbrand G, Perrin I, Oswald IP, Speijers G, Chiodini A, Recker T, Dussort P (2016) Impact of food processing and detoxification treatments on mycotoxin contamination. Mycotoxin Res 32(4):179–205. https://doi.org/10.1007/s12550-016-0257-7

Kolosova A, Stroka J (2011) Substances for reduction of the contamination of feed by mycotoxins: a review. World Mycotoxin J 4:225–256

Kumari A, Singh K (2021) Evaluation of prophylactic efficacy of cinnamaldehyde in murine model against *Paradendryphiella arenariae* mycotoxin tenuazonic acid-induced oxidative stress and organ toxicity. Sci Rep 11:19420. https://doi.org/10.1038/s41598-021-98319-8

Lopez-Garcia R, Park DL (1998) Effectiveness of post-harvest procedures in management of mycotoxin hazards. In: Bhatnagar D, Sinha S (eds) Mycotoxins in agriculture and food safety. Marcel Dekker, New York, pp 407–433

Lopez-Garcia R, Park DL, Phillips TD (1999) Integrated mycotoxin management systems. Food, nutrition and agriculture. http://www.fao.org/docrep/X2100T/x2100t07.htm

Magan N, Medina A, Aldred D (2011) Possible climate-change effects on mycotoxin contamination of food crops pre-and post harvest. Plant Pathol 60(1):150–163. https://doi.org/10.1111/j.1365-3059.2010.02412.x

Pankaj SK, Shi H, Keener KM (2018) A review of novel physical and chemical decontamination technologies for aflatoxin in food. Trends Food Sci Technol 71:73–83

Schofield EK (2005) Feed manufacturing technology. American Feed Industry Association, Inc., Arlington

Shephard GS, Burger HM, Rheeder JP, Alberts JF, Gelderblom WC (2019) The effectiveness of regulatory maximum levels for fumonisin mycotoxins in commercial and subsistence maize crops in South Africa. Food Control 97:77–80

Sorrenti V, Di Giacomo C, Acquaviva R, Barbagallo I, Bognanno M, Galvano F (2013) Toxicity of ochratoxin A and its modulation by antioxidants: a review. Toxins 5:1742–1766

Stangroom KE, Smith TK (1984) Effect of whole and fractionated dietary alfalfa meal on zearalenone toxicosis and metabolism in rats and swine. Can J Physiol Pharmacol 62:1219–1224

Thio S, Wijaya S (2010) The concern and awareness of consumers and food service operators towards food safety and food hygiene in small and medium restaurants in Surabaya, Indonesia. Int Food Res J 17:641–650

Thomas M, Hendriks WH, van der Poel AFB (2018) Size distribution analysis of wheat, maize and soybeans and energy efficiency using different methods for coarse grinding. Anim Feed Sci Technol 240:11–21

Vanara F, Scarpino V, Blandino M (2018) Fumonisin distribution in maize dry-milling products and by-products: impact of two industrial degermination systems. Toxins (Basel) 10(9):357. https://doi.org/10.3390/toxins10090357

van Egmond HP, Schothorst RC, Jonker MA (2007) Regulations relating to mycotoxins in food: perspectives in a global and European context. Anal Bioanal Chem 389:147–157. https://doi.org/10.1007/s00216-007-1317-9

Vidosavljević S, Kos J, Banjac V, Janić Hajnal E, Dragojlović D, Uragić O, Ćolović R (2018) Simple technologies for removal of mycotoxins in feed. In: Proceedings of the feed technology, vol 18. University of NoviSad, Institute of Food Technology, Novi Sad, pp 81–88

Whitlow LW, Hagler Jr WM (2006) Mycotoxins, cattle and dairy food products. The top ten most frequently-asked questions about mycotoxins, cattle and dairy food products. https://en.engormix.com/mycotoxins/articles/mycotoxins-cattle-dairy-foodproducts-t33449.htm

WHO (2018) Mycotoxins. https://www.who.int/news-room/fact-sheets/detail/mycotoxins

Chapter 10
Toxicokinetics of Mycotoxins

10.1 Introduction

Toxicokinetics and toxicodynamics are two parameters that determine the toxicity of a compound. Toxicodynamics is the interaction of the substance with biological targets and their downstream biological effects. On the other hand, the relationship between the concentration of a particular substance to which a human being is exposed to, and the concentration of the chemical that is toxicologically active at the point of action which might be an organ, tissue, cell, or molecules where the compound shows its effect, is known as toxicokinetics (DellAfiora et al. 2018).

10.2 Routes of Exposure and LADME

Skin absorption, inhalation, and oral intake are the various routes of exposure of mycotoxins in humans and animals (DellAfiora et al. 2018). The concept of LADME describes the difference in the concentration of the ingested compound and its concentration at the site of action. More precisely, the intensity of the effect of any toxicant that is consumed along with food is dependent on its bioaccessibility, a factor that determines the part of toxicants that is released from the food in the gastrointestinal tract on digestion, thus becoming available for the absorption (L = liberation). A part of the toxicants that are released are absorbed (A = absorption) and reach the bloodstream from where they are distributed into the body tissues (D = distribution). A number of metabolites are then released (M = metabolism) at both pre-systemic and systemic levels before they are eliminated through the biliary or renal routes (E = excretion) (DellAfiora et al. 2018).

One of the most important determinants for toxicity is exposure and, once absorbed into the systemic circulation, the ability of the organism to effectively eliminate the toxic compound from the body. The above aspects are responsible for

K. Singh, A. Kumari, *Mycotoxins and Mycotoxicoses*,
https://doi.org/10.1007/978-981-19-2370-8_10

the variations in sensitivity of a species towards a particular toxicant. For instance, initially rapid conversion of ZEN to α-zearalenol (α-ZEL) occurs in pigs, whereas in chickens it is transformed to β-zearalenol (β-ZEL). α-ZEL binds more easily to oestrogen receptors than ZEN and β-ZEL as α-ZEL has a better binding affinity towards oestrogen receptors than the other two. This is the reason why pigs are more sensitive to ZEN. In case of DON, the capacity of detoxification has a difference similar to that of ZEN, which explains the reason that ruminants are less sensitive to the effects of DON. In the rumen of healthy cattle, DON is converted to its de-epoxy form at once. Conversely in pigs, DON is almost completely and very quickly absorbed in the first part of the intestine while the process of de-epoxidation takes place in the distal intestine, thereby not making de-epoxidation a noteworthy detoxification pathway in pigs. This is why pigs have a higher sensitivity towards DON (Schelstraete et al. 2020).

10.3 Toxicokinetics of Some Important Mycotoxins

10.3.1 Aflatoxin and OTA

In 2012, toxicokinetic studies on co-administration of aflatoxin B1 and ochratoxin A to rats were carried out by Corcuera et al. F344 rats were administered with 0.25 mg/kg bw Af B1 and 0.5 mg/kg bw OTA. At different time points, blood, kidney, and liver were collected for quantification of mycotoxin, calculation of weight, biochemical, and histopathological studies.

Af B1 resulted in acute toxicity in the liver only. It reached a maximum concentration (C_{max}) in plasma at 10 min while OTA reached the maximum concentration at 2 h. Levels of Af B1 in blood showed quick absorption and its metabolism; however, its concentration in the kidney was lower than LOD and in the liver was lower than LOQ. C_{max} for OTA in the plasma was recorded to be 4326.2 μg/L and in the kidney and liver C_{max} was attained at 8 h and the concentration was almost equal in the two organs (Corcuera et al. 2012).

10.3.2 Aflatoxins

10.3.2.1 Biodistribution, Metabolism, and Excretion

Metabolism of aflatoxins B1, G1 and other carcinogenic compounds takes place in the liver and other organs in phases namely phase 1 and phase 2. In mammals, phase 1 metabolism is carried out by cytochromes P450 (CYP) while in humans CYP1A2, CYP3A5, and CYP3A4 are involved in the phase 1 process. Metabolism in phase 1 lowers the toxicity of the molecule and results in the formation of a more reactive electrophilic species that readily combines with macromolecules of the cell. For instance, aflatoxin M1, a product of aflatoxin B1 phase 1 metabolism catalysed by CYP1A2 is around ten times less powerful a carcinogen than aflatoxin B1. Others include aflatoxins P1 and Q1 with lower toxicity after metabolism (Haschek et al. 2013).

Bioactivation of Af B1 occurs by the conversion of Af B1 to Af B1-8,9-epoxide and other epoxides. These epoxides additionally go through phase 2 glutathione-*S*-transferase-mediated conjugation to glutathione following exertion through urine in the form of aflatoxin-mercapturic acid. They are also converted to dihydrodiols or dialdehydes and bind covalently to macromolecules. Aflatoxin epoxide binds to DNA particularly at the N7 sites of guanine nucleosides resulting in the formation of aflatoxin–N7 guanine adduct. Aflatoxin–N7 guanine adduct, a product of depurination, is formed during the DNA repair process. This adduct is present in the urine and blood and has proven to be very advantageous as a biomarker in epidemiological studies in detecting recent exposures to the toxin. Mycotoxin detection in the body becomes easier with the formation of aflatoxin–albumin adduct in serum and aflatoxin-mercapturic acids in the urine (Deng et al. 2018; Bbosa et al. 2013; Haschek et al. 2013).

Depending on the type and quantity of metabolites formed in phase 1 and phase 2 of metabolism, variations in the susceptibility of aflatoxicosis among different species and different sex are determined. An example for this is the elevated levels of glutathione-*S*-transferase activity in adult mice as compared to that in rats and other sensitive species that cause relative insensitivity to hepatocarcinogenicity. The phase 2 metabolism of aflatoxin in the cytosol is not as much efficient in humans as in mice (Deng et al. 2018; Bbosa et al. 2013; Haschek et al. 2013).

10.3.2.2 Mechanism of Action

Exposure to aflatoxins results in hepatotoxicity. The mechanism of action of aflatoxins and their metabolites is extensive and not specific. They interact readily with several cell proteins and leads to the disruption of the basic metabolic processes and protein synthesis leading to the death of the cell (Haschek et al. 2013).

Aflatoxins also possess genotoxic property. The formation of N7 guanine-aflatoxin adducts in the urine in correlation with exposure to aflatoxin has been demonstrated frequently. The addition reaction of aflatoxin to DNA results in mutation. These mutations occur while DNA repair and replication; and when critical genes are involved, changes in the function of the cell occur (Haschek et al. 2013). For instance, exposure to aflatoxin causes a point mutation at the third base of codon 249 of the TP53 tumour suppressor gene. Commonly reported in liver cancer patients in China and Africa, this mutation causes guanine transversion to thymidine (AGG to AGT). Though, some studies do not agree with the correlation between aflatoxin carcinogenesis and TP53 mutations (Deng et al. 2018; Bbosa et al. 2013; Rushing and Selim 2019).

10.3.2.3 Aflatoxin in Pregnant Mice

Toxicokinetics of Af B1 was studied in pregnant mice by dosing the mice on day 13 of gestation. Af B1 was administered intra-peritoneally (IP) and orally at a dose of 20 mg/kg to pregnant mice. Control mice were administered with a proportionate quantity of solvent. Blood was collected at 15 min, 30 min, 45 min, 1 h, 1 h 30 min, 2 h, and 1 h 30 min. Af B1 concentration was analysed in the serum and it was noted that C_{max} was attained at 15 min of IP and 30 min of oral dose. Aflatoxins were quickly and easily absorbed in the gastrointestinal tract (GI) tract and through the peritoneum. The absorption rate of Af B1 was 5.0 μg/min and the elimination rate was 3.0 μg/min (Bastaki et al. 2010).

10.3.3 Deoxynivalenol

Depending on geographical location, the predominant trichothecenes may be DON, nivalenol (NIV), or one of their acetylated precursors. However, DON is the focus of this section (Sugita-Konishi and Nakajima 2009; Pestka 2010a, b; http://publications.iarc.fr/_publications/media/download/1375/23128bfd0453d69904d218c8aa9b9bf5480fc110.pdf).

10.3.3.1 Absorption

DON is quickly absorbed, distributed, and eliminated after either oral or parenteral dosing. DON is not accumulated in tissues or transmitted to eggs or milk at levels in which it is consumed (Prelusky et al. 1994). It is extensively metabolized in the gastrointestinal tract by the process of de-epoxidation resulting in the formation of de-epoxy metabolite of DON (DOM-1) (Pestka 2010a, b; WHO 2001).

10.3.3.2 Gastrointestinal Metabolism

Microorganisms in the gastrointestinal tract have frequently been reported to metabolize DON. Substantial conversion of DON to the de-epoxy metabolite (DOM-1) has been illustrated by incubation of DON with gastrointestinal tract cultures or extracts (WHO 2001). Pigs, chickens, and bovine rumen fluid show almost complete de-epoxidation. Pigs incapable of carrying out intestinal de-epoxidation can start de-epoxidation by simply coming in contact with faeces of pigs that have the capacity to (Eriksen et al. 2003). 3-Acetyl DON is de-acetylated to DON in the gastrointestinal tract of pigs before it gets absorbed. De-epoxides of DON are not as cytotoxic as the parent compounds. Unlike de-epoxidation, de-acetylation takes place in the distal parts of the digestive tract where absorption of DON is low (Eriksen et al. 2003).

10.3.3.3 Bioavailability

Studies show that absorption of DON begins in the stomach or upper duodenum as its maximum concentration in plasma reaches within 2.8 h of feeding pigs. DON is detected in the plasma within 20 min of feeding with 3-acetyl DON for 3 days (Eriksen et al. 2003). Half-life of DON is 3.9 h in pigs and 100–125 min in sheep after intravenous injection. A single intra-gastric injection in pigs showed that the peak of DON in plasma was recorded after 15–30 min with half-life of 7.1 h. Bioavailability of DON is low in cows, 55% in pigs and <10% in sheep. Sheep, pigs, and cows show quick absorption of DON in the gastrointestinal tract (WHO 2001).

10.3.3.4 Distribution

Studies on the distribution of DON suggest that DON is distributed quickly but does not accumulate in tissues. Radiolabelled DON is available within 3 h of oral administration in chickens and its distribution is bile ≫ kidney > blood/plasma > liver ≫ other tissues (WHO 2001). After 72 and 96 h, DON was either not present or present in a very small quantity. After intravenous administration in pigs, relative accumulation of DON was kidney > plasma > liver > fat ≫ other tissues (WHO 2001). The level of DON reduces by 90% after 24 h of dosing as compared to that after 3 h of dosing (WHO 2001).

10.3.3.5 Excretion

Quick elimination of DON occurs in sheep, pigs, and chickens through urine and faeces (WHO 2001). After 24, 48, and 72 h of oral dosing, 79%, 92%, and 98% DON were recovered in the excreta of chickens, respectively. 6.9%, 0.11%, and 64% of the dose were encountered in the urine, bile, and faeces, respectively, in sheep. In pigs and ewes, after 24 h of intravenous injection, 91% and 6% was present in the urine and bile, respectively. Fifty to 62% of DON is excreted in the urine in pigs that feed on naturally contaminated wheat (Goyarts and Dänicke 2006). When 3-acetyl DON was fed to pigs, 2% and 45% of the dose were recovered in faeces and urine, respectively, after 48 h when the pigs are taken off the diet (Eriksen et al. 2003).

10.3.3.6 Transmission

Transmission of DON to edible animal products is minute and thus poses little health risk to consumers. In sheep and cow milk, transmission of DON is less. When chickens are fed with feed contaminated with DON for 65 days at a concentration of 5.5 mg/kg diet, very low levels (<1.7 mg/egg) of DON is detected in eggs. DON has not been detected in the milk of cows consuming naturally contaminated maize with DON (12 mg/kg) for 10 weeks (WHO 2001).

10.3.3.7 Metabolism

Metabolism of DON by microsomal enzymes has not been recorded yet. However, the presence of glucuronide conjugation in pigs and sheep has been demonstrated (WHO 2001; Eriksen et al. 2003). Studies showed that 42% and 33% of the DON were in plasma and urine, respectively, when pigs were fed with feed containing 3-acetyl DON. A significant percentage of DON was also present in the form of de-epoxides in faeces (Eriksen et al. 2003).

10.3.4 Ochratoxin A

10.3.4.1 Absorption

It has been reported that different percentage of OTA is absorbed in different species. The percentage absorption in pigs is 66%, in rats and rabbits is 56%, and in chicken is 40% (Hadjeba-Medjdoub et al. 2012). Adsorption of OTA is greatly dependent on the pKa of its carboxyl group of the phenylalanine moiety and the phenolic hydroxyl group of the isocoumarin part. In the upper parts of the GI tract, OTA is mostly present in the protonated form (OTA^{+}) while in the duodenum where the pH is physiologically normal, it is present in the monoanion (OTA^{-}) and the dianion (OTA^{2-}) form. Passive absorption of OTA takes place from the stomach and the proximal jejunum mostly in the non-ionized (OTA^{0}) and monoanion (OTA^{-}) forms (Manderville 2005; Turesky 2005; Manderville and Pfohl-Leszkowicz 2008; Malir et al. 2014). OTA readily binds to plasma proteins thus favouring its passive absorption. To determine the mechanism of OTA absorption, in vitro studies on human intestinal epithelial cells (CaCo-2) have been carried out (Pfohl-Leszkowicz and Manderville 2012). After absorption, OTA is transported and distributed to various tissues and organs by the portal system (Kőszegi and Poór 2016).

10.3.4.2 Distribution

Binding to Plasma Proteins

OTA has a high affinity for serum proteins especially albumin and as OTA reaches the circulation, 99% of it is absorbed via passive absorption in the non-ionized form resulting in a prolonged half-life (El Khoury et al. 2010; WHO 1993). Only traces of OTA are found in erythrocytes (Manderville 2005).

Binding of OTA to albumin delays its elimination as its transfer to the hepatic and renal systems is reduced. Studies carried out in this context show that rats that were deficient in OTA were able to remove OTA from the systemic circulation about 20- to 70-fold more rapidly as compared to control rats (National Toxicology Program 1989).

In another study carried out by Stojkovic et al. (Cui et al. 2010), it was found that OTA could bind more specifically to two small serum proteins (around 2 kDa) as compared to albumin. These two molecules being smaller in size could easily cross the glomerular membrane. When OTA is bound to these molecules, it crosses the glomerular membrane with the molecule thus rendering OTA's nephrotoxic effect in mammals. Binding saturation of OTA in serum albumin is more than several hundred μg/mL while the specific binding saturation on these molecules was achieved at an OTA concentration of 10–20 ng/mL serum. Some other blood proteins have also been reported in the transportation of OTA (Liu et al. 2015).

In experiments carried out by Hagelberg et al. (Schwartz 2002), toxicokinetics of OTA in several animals were studied. It was found that unbound percentage of OTA was 0.02% in humans and rats, 0.08% in monkeys, 0.1% in mice and pigs, and 22% in fish when administered at a concentration of 50 ng OTA/g body weight. The OTA bound to the serum proteins are present in reserve and is liberated to the tissue. The half-life of OTA in vervet monkey was found to be 456–504 h when administered intravenously (Vidal et al. 2015; Pohland et al. 1982). The maximum serum half-life of the mycotoxin has been recorded in humans. OTA has a longer half-life when administered intravenously as compared to that of oral administration. This might be because before entering the systemic blood circulation, OTA might encounter transit through the liver and then elimination by the bile.

Tissular Distribution

The concentration of toxin and its metabolites present in plasma, tissues, and organs of animals depend on several factors like animal species, quantity of toxin administered, the form of the OTA (crystalline or naturally occurring in feed), the health status, and diet composition of the animal. The presence of toxin residues in animal tissues is greatly influenced by these factors. The half-life of OTA in blood is more than in tissues, this is due to the fact that OTA binds to the blood proteins more rapidly than to tissues (Schwartz 2002, Trivedi et al. 1992; Van der Stegen et al.

2001; Cavret and Lecoeur 2006; Asam et al. 2013). When administered in low quantities, distribution of OTA in rats was in the order: lung > liver > kidney > heart > fat > intestine > testes > muscle > spleen > brain (Chen et al. 2015). In rats, chickens, goats, and pigs, maximum distribution of OTA was seen in the kidney, liver, muscle, and then in fat (decreasing order) (Malir et al. 2014; Flores-Flores et al. 2015; Shim et al. 2014; Veprikova et al. 2015).

Primary cell cultures from different organs of rat and various cell lines of kidney were used by Schwerdt et al. (Liu et al. 2015) to check the specific binding capacity of OTA to various proteins of the cytosol and cell organelles. OTA showed apparent binding with a scarce organellar protein (62 kDa) present in the cells arising from the proximal tubules. This might be another factor contributing to the nephrotoxic effect of OTA by the accumulation of OTA in renal cells.

When 289 g OTA/kg BW/day was administered to male rats via gastric intubation for 8 days, 0.022–0.028% of the dose was found in the hippocampus, striatum, ventral mesencephalon, and brain of the rats (Veprikova et al. 2015).

When high concentrations of radiolabelled OTA was administered to mice, rats, and swine intravenously; it was found that at the eighth and ninth day of gestation, placental crossing of OTA was higher than the tenth day. Within 20 min of crossing the uterine wall, OTA was detected in the placenta and foetal tissues (Ostry et al. 2013). The differences noted in the transfer of OTA in the foetus at different days of gestation show that the foetal uptake of OTA is governed by the stages of development of the placenta which develops completely after the ninth day of gestation (Kőszegi and Poór 2016).

10.3.4.3 Elimination

OTA is absorbed quickly and eliminated slowly. It is cleared from plasma via urine or faecal matter in all species (Pavlović et al. 1979; Eriksen et al. 2003; Duarte et al. 2009). Several factors including the dose, the route of administration, capacity to bind with plasma proteins, and enterohepatic circulation contribute towards the excretion of OTA. The mammalian milk excretion, urinary, and faecal routes are important routes of elimination of OTA (Kőszegi and Poór 2016).

10.3.5 Alternaria *Mycotoxins: AOH, AME, and TeA*

The total body clearance (Cl) of TeA in pigs (0.45 L/h/kg) and broiler chickens (0.06 L/h/kg) is quite low. The low volume of distribution (V_d) in both broiler chickens (0.2 L/kg) and pigs (0.3 L/kg) indicate that the distribution of the toxin to the tissues is limited. In broiler chickens, the low Cl could perhaps result in the accumulation of TeA when they are fed with contaminated diet on a daily basis, leading to the deterioration of the health of the animal and increasing the threat of carry-over of the mycotoxin in the food chain. Thus, more studies based on tissue residue are recommended. TeA has high bioavailability in humans when consumed orally as nearly 90% of the ingested TeA can be recovered in the urine (Fraeyman et al. 2015). Recent studies carried out on toxicokinetics of AOH showed that 90% of the total dose administered was eliminated through faeces and almost 9% through urine thereby, suggesting low systemic absorption of AOH (Asam et al. 2013; Schuchardt et al. 2014; Aupanun et al. 2016).

10.3.5.1 Alternariol (AOH) and Alternariol Monomethyl Ether (AME)

Studies on *Alternaria* toxins with dibenzo-pyrone structure (AOH and AME) and perylene quinone (ATXs, ALTCH, and STEs) structures have been carried out; however, not much data is available on the toxicokinetics of ATXs, ALTCH, and STEs. Studies were carried out to check the absorption and metabolism of four *Alternaria* toxins with perylenequinone structures. Absorption in the intestine and metabolism were checked in vitro using the Caco-2 cell transwell system. Results showed that ATX I and ALTCH were not metabolized in Caco-2 cells but ATX II and STE III were biotransformed partially by reductive de-epoxidation to ATX I and ALTCH metabolites, respectively (Fleck et al. 2014). ATX-II showed very little absorption and was metabolized partially in the intestinal Caco-2 transwell system (Pahlke et al. 2016).

Toxicokinetic studies on TeA showed complete oral bioavailability in swine and broiler. TeA absorption was slow in broiler chickens (2.60 h) as compared to pigs (0.32 h) while elimination was slower in chickens with $t_{1/2}$ 2.45 h in broilers and $t_{1/2}$ 0.55 h in pigs post oral administration. Total body clearance in pigs was 446.1 mL/h/kg and in broilers was 59.2 mL/h/kg after oral administration (Asam et al. 2013). Human exposure of TeA showed that 87–93% of TeA ingested was eliminated as a parent compound in the urine after 24 h of consumption in two volunteers. In six volunteers, TeA was present in the urine at a concentration of 1.3–17.3 μg/L (Fraeyman et al. 2015; Escrivá et al. 2017).

10.3.5.2 Tenuazonic Acid (TeA)

A single per os (po) and intravenous (iv) dose of TeA at a concentration of 0.05 mg/kg BW showed no side effects. After iv and po administration of TeA in chickens, mean AUC_{0-t} values were 805.6 and 893.2 ng/h/mL while that in pigs were 108.9 and 151.7 ng/h/mL. Complete bioavailability of TeA after oral administration in pigs (mean $F = 1.47$) and broiler chickens (mean $F = 1.24$) was noted. The highest mean concentration of TeA was reached after 32 h in pigs and at 2.60 h in broiler chickens. Rapid elimination of TeA in pigs and broilers takes place after oral dosing. TeA has a half-life of 0.51 h after iv administration and 0.55 h after po administration in pigs while in broiler chickens, it has a half-life of 2.03 h after iv and 2.45 h after po administration. In pigs, the mean clearance value is 448.4 mL/h/kg post iv injection and 446.1 mL/h/kg post po dosing while that in broilers the mean Cl values are

59.3 mL/h/kg post iv administration and 59.2 mL/h/kg post po dosing. The mean volume of distribution in pigs was 325.8 and 358.4 mL/kg when administered through iv and po routes, respectively, and was 164.6 and 207.2 mL/kg in broilers via iv and po routes, respectively. In broiler chickens, the mean MRT values were 4.46 and 5.52 h while in pigs, it was 0.70 and 0.94 h after iv and po administration, respectively. The difference in plasma protein binding in pigs was 58.1 ± 9.2% and in broilers was 72.8 ± 3.4 which was not significantly different ($p = 0.06$) (Fraeyman et al. 2015).

10.3.6 *Fusarenon-X (a Type B Trichothecene Mycotoxin)*

10.3.6.1 Absorption and Distribution

In broilers, ducks, mice, and piglets, FX is rapidly absorbed in the gastrointestinal tract as it is highly lipid-soluble. FX reaches maximum concentration in the plasma after 5 min in piglets, after 12 min in ducks and after 30 min in mice. In mice, oral bioavailability of FX is higher than NIV, its metabolite. Piglets (74%) showed highest oral bioavailability as compared to ducks (19.5%) and broilers (9.8%) (Poapolathep et al. 2003, 2008; Cavret and Lecoeur 2006; Saengtienchai et al. 2014; Aupanun et al. 2016).

10.3.6.2 Metabolism

After 3 h of oral administration, the maximum level of FX was noted in the liver, kidney, and spleen of piglets (Ohta et al. 1978). NIV (FX metabolite) was present only after 20 min of oral exposure in the plasma of ducks and broilers (Saengtienchai et al. 2014). In vitro studies also showed that microsomal carboxyesterase hydrolyses FX to NIV. Thus, studies reveal that FX was quickly metabolized to NIV in the liver and kidney after being absorbed in the gastrointestinal tract (Poapolathep et al. 2004). Cell culture studies on microsomal nonspecific carboxyesterase in rats and rabbits show that the C-4 acetyl residues of FX are hydrolysed to NIV via microsomal carboxyesterase. The results of these studies suggest that FX is metabolized to NIV as soon as it is absorbed in the gastrointestinal tract. In vitro studies on metabolism of FX showed that FX is converted to NIV in the liver and kidney. Further, in vivo studies on mice, piglets, broilers, and ducks also confirm that FX is converted to NIV mainly in the liver and kidney (Poapolathep et al. 2003, 2004; Saengtienchai et al. 2014; Ohta et al. 1978). In mice, conversion percentage was more in the liver, i.e., 93.99% while that in the kidney was 27.91%. In piglets and ducks, the conversion percentage was almost the same. In ducks, it was 98.95% in the liver and 94.32% in the kidney while in piglets it was 90.91% in the liver and 89.72% in the kidney. This conversion was the opposite in the case of broilers where the conversion percentage in the liver was 70.12% and 94.39% in the kidney

(Poapolathep et al. 2003; Saengtienchai et al. 2014; Ohta et al. 1978). An interesting result showed that NIV was also found in foetal mice through the placenta and suckling mice through the mother's milk after FX was metabolized to NIV in the mother's body (Wu et al. 2010). Furthermore, the metabolism of NIV to its de-epoxidated form is brought about by the microorganisms of the gastrointestinal tract (Kongkapan et al. 2015). The intestinal microflora plays an important role in the biotransformation of trichothecenes. They convert NIV to less toxic compounds. The presence or absence of a particular microorganism controls the sensitivity of an animal towards NIV thus making intestinal microflora an influential factor in toxicity (Norred et al. 1985).

10.3.6.3 Elimination

Intravenous and oral administration of FX in piglets show that FX and NIV are present in the urine and faeces for up to 24 and 48 h, respectively. Exposure to FX resulted in large amounts of NIV in the urine (Ohta et al. 1978). Faeces of broilers and ducks also showed the presence of FX and NIV after iv and po administration (Poapolathep et al. 2003). An early study suggested that administering ^{3}H-FX to mice showed high NIV concentration in the urine and lower FX concentration (Saengtienchai et al. 2014). Similar was the pattern for mice excreta also. Therefore, it was inferred that within 24 h of administration, FX is either excreted out quickly or transformed almost completely to NIV and eliminated through urine. The tissues of mice namely liver, lung, kidney, spleen, stomach, heart, brain, thymus, mesenteric lymph nodes, bone marrow, small and large intestine, caecum, muscle, and skin showed similar concentrations of FX. It could thus be concluded that the accumulation of FX in the spleen, bone marrow, mesenteric lymph nodes, and thymus depends on the metabolites of FX along with the concentration of FX administered (Saengtienchai et al. 2014; Kongkapan et al. 2015; Norred et al. 1985; Aupanun et al. 2016).

10.3.7 Cyclopiazonic Acid

10.3.7.1 Absorption

In experiments carried out by Norred et al. (1985, 1990), radioactive CPA was injected intra-gastrically (5 mg/kg) and intra-peritoneally to Sprague-Dawley rats (Byrem et al. 1999a; Norred 1990). Their excreta were collected at intervals of 2, 3, 6, 12, and 24 h and tissues were collected after 72 h. After 1 h of IP administration of the mycotoxin, a peak level of 14.5% CPA was detected and 10% CPA was detected after 6 h. A constant decrease in the level of CPA was noted from 6 to 72 h regardless of the route of dosing. In pigs, an average plasma concentration of 464 ng/mL was detected when they were fed with a diet containing 10 ppm of CPA for 6 days. The concentration of CPA in muscles was similar to that in plasma (Byrem et al. 1999b).

10.3.7.2 Distribution, Metabolism, and Elimination

The concentration of CPA reached peak levels in the blood after 6 h of intra-gastric and 1 h of IP dosing. The highest amount of CPA was found in blood and then in the organs such as the liver, lung, kidney, heart, and skeletal muscles. The tissues in which the specific activity of CPA was recorded to be minimum in both intra-gastric and IP routes were the spleen, brain, testis, eye, thyroid, adrenal gland, salivary gland, fat, and skin. After 1 h of IP administration of radioactive CPA, its rapid accumulation was noted in the liver of rats. However, after 1 h, the levels were lowered as compared to rats with intra-gastric administration. Further, CPA concentration was lowered to $<1\%$ of the dose after 72 h in both the routes of dosing. The maximum level of CPA was found in the skeletal muscles at all the time intervals. Accumulation in muscle was more in the IP route but from 6 to 72 h distribution was directly proportional to time and was similar for both the routes. Around 45–50% of the CPA or its metabolites were distributed to muscles during the first 12 h of dosing (Norred et al. 1985; Norred 1990). The accumulation of CPA in muscles is a reason for the clinical symptoms including lethargy, incoordination, and weakness (Byrem et al. 1999a; Norred 1990).

A three-component model for the distribution of CPA in pigs was studied which showed that when 20 mg CPA was injected iv to pigs, it was distributed rapidly. The volume of distribution was large and elimination occurred slowly (Byrem et al. 1999b). The concentration of CPA in muscle and plasma were equivalent when pigs were fed ad libitum with feed containing 10 ppm CPA which was almost equal to that of the total water content in the body of pigs (Byrem et al. 1999b).

In rats, IP and IG dosing of rats showed that a significant amount of CPA was eliminated, i.e., $86.9 \pm 10.1\%$ (IP) and $95 \pm 10.9\%$ (IG). The half-life of CPA was 33 ± 12 h in the IP route while that for the IG route was 43 ± 25 h. Twenty-three per cent of CPA administered was eliminated via urine and 37% via faeces in the IP route, 72 h post dosing while 17% was excreted in the urine and 48% in the faeces in IG route in the same time period. Expired air did not contain any radioactivity. When administered intra-peritoneally, 5% of the dose was noted in the GI tract after 1 h of dosing which increased to 13%, 12 h post dosing and in the next 60 h, came down to 5–9%. On the other hand, on GI dosing, radioactive CPA decreased rapidly for the first 12 h and then came down slowly to 3% up to the next 72 h (Norred et al. 1985; Norred 1990).

In pigs, the half-life ($t_{1/2}$) of CPA was long, i.e., 24 h after a single iv dose of 20 mg. However, the route of elimination in pigs still needs to be studied (Burdock and Flamm 2000; Norred et al. 1985; Byrem et al. 1999a). In experiments carried out by Byrem et al. (1999a, b), four pigs were administered with a single injection of CPA containing 20 mg CPA. Plasma was collected from the animal from 2 to 96 h. Another group of 3 pigs were fed a diet containing 10 mg CPA/kg bw for 6 days.

Collection of blood samples was then performed at the third, fourth, and sixth day of the dosing. At the sixth day, skeletal muscles were also collected. A three-compartment model was used to describe the CPA plasma kinetics. It was observed that distribution of CPA was quick and V_d was large, i.e., 49 L, when 20 mg CPA was administered through the intravenous route. CPA level became steady within 6 days of dosing, when fed with 10 mg CPA/kg bw and had a *t*½ of 24 h. The concentration of CPA in skeletal muscles was 469 ± 86 ng/g and plasma was 410 ± 44 ng/mL (Byrem et al. 1999a, b). In another experiment, radioactive CPA [^{12}C] was administered via three different routes parenterally (1 mg/kg, or 0.12 μCi/kg), IG (5 mg/kg, 0.6 μCi/kg), and orally (5 mg/kg, 0.6 μCi/kg) to Spargue-Dawley rats. Depending on the route of administration, $t_{½}$ of CPA varied from 33 to 43 h. The maximum level of CPA was reached in the GI tract within 6 h of dosing (10% of dose) and the liver, lungs, kidney, and heart showed high concentrations of labelled CPA. Muscle tissue showed the presence of 45–50% of dosed CPA in the first 12 h of dosing. Urine and faeces formed the major route of excretion of CPA and its metabolites. CPA showed a dose-dependent relationship in chickens, and the highest level of CPA was detected in its meat 3 h of post dosing when 0, 0.5, 5.0, or 10 mg/kg bw of CPA was administered via intubation. CPA percentage decreased in muscles soon after 3 h when low and intermediate doses of CPA were dosed in birds. No CPA was found in muscles after 24 h of dosing when low doses of CPA were given and at 3 h, 25% of the concentration of CPA was found when moderate levels of CPA were dosed. The biological *t*½ of CPA in the bird was 60 h and its elimination rate from muscles was slower. CPA is rapidly absorbed and distributed in the tissue with the highest concentration in skeletal muscles in rodents (Malekinejad et al. 2011).

10.3.8 Beauvericin and Enniatins

Swine showed high oral bioavailability ($F = 90.9\%$) of ennianitin B1 (ENN B1) while in broilers, ENN B1 and ennianitin B (ENN B) were not readily absorbed with an *F* of 5% and 11% for ENN B1 and ENN B, respectively (Devreese et al. 2014; Fraeyman et al. 2016). On comparison, broilers show higher body clearance (Cl) and volume of distribution (V_d) of ENN B1 than swine. IV administration of ENN B1 showed a Cl of 6.6 L/h/kg in broiler chickens and 1.91 L/h/kg in swine. Broiler chicken showed a V_d of 25 L/kg for ENN B1 (EFSA 2014). On the other hand, swine showed a V_d of 0.57 and 0.69 L/kg for the central and peripheral compartment, respectively, in a two-compartmental model.

Broilers and swine showed the presence of hydroxylated and carboxylated metabolites of ENN B and B1 in plasma. Swine also showed the presence of carbonylated metabolites of ENN B1 in plasma. A higher concentration of ENN B1 metabolite in swine after oral dosing in comparison to that after iv dosing signifies that pre-systemic metabolism of ENN B1 takes place. In vitro studies documented that ENNs and beauvericin (BEA) alter the barrier integrity of the intestine thus, not only affecting the bioavailability of xenobiotics but also making it more susceptible to diseases and reduced performance. Altogether, these mycotoxins affect different species differently and more species and mycotoxin specific studies are needed so as to attribute their toxicokinetic behaviour and differences (Devreese et al. 2014; Fraeyman et al. 2016).

Very limited studies have been carried out regarding the biotransformation of BEA and ENNs in the food of animal origin. Contamination of BEA and ENNs has been found in eggs (56–99.7%). Low levels of co-contamination with a minimum of two toxins have been seen in yolk samples. Biotransformation of ENN A and ENN A1 in traces has been noted only in the meat of chickens while ENN B, ENN B1, and BEA have been found in their meat, skin, and liver. Percentage carry-over rate of ENN B was maximum in the skin, 0.39%; 0.16% in liver, 0.04% in thigh muscle and 0.01% in breast muscle; while that for ENN B1 was 0.37% in the skin, 0.16% in liver, 0.04% in thigh muscle, and 0.025% in breast muscle. Highest concentration of BEA was present in liver (1.57%) > skin (1.16%) > muscle (0.03%). Liver and meat show complete elimination of ENNs and BEA while in the case of skin elimination is incomplete and slow. The presence of ENN B, B1, and BEA was identified in eggs after 2–3 days of consumption of contaminated feed and was highest at day 5. Complete elimination of ENNs and BEA from the eggs took 10 days during which they received a control diet. Percentage carry-over rate in eggs was highest in BEA (0.44%) > ENN B (0.10%) > ENN B1 (0.05%) (Callebaut et al. 2006; Jestoi et al. 2009; Springler et al. 2016; Schelstraete et al. 2020; Zhao et al. 2015).

10.3.9 Moniliformin

Not many studies have been carried out on the toxicokinetics of moniliformin (MON). In vitro studies prove that MON can pass through the blood-brain barrier but no affect was found in the intestinal or blood capillary integrity. Only one oral dose of MON showed that 42% of MON was eliminated via urine while >1% was excreted through faeces within 24 h of dosing (Schelstraete et al. 2020; Jonsson et al. 2013; Behrens et al. 2015).

References

Asam S, Habler K, Rychlik M (2013) Determination of tenuazonic acid in human urine by means of a stable isotope dilution assay. Anal Bioanal Chem 405(12):4149–4158. https://doi.org/10.1007/s00216-013-6793-5

Aupanun S, Poapolathep S, Giorgi M, Imsilp K, Poapolathep A (2016) An overview of the toxicology and toxicokinetics of fusarenon-X, a type B trichothecene mycotoxin. J Vet Med Sci 16:0008. https://doi.org/10.1292/jvms.16-0008

Bastaki SA, Osman N, Kochiyil J, ShAfiullah M, Padmanabhan R, Abdulrazzaq YM (2010) Toxicokinetics of aflatoxin in pregnant mice. Int J Toxicol 29(4):425–431. https://doi.org/10.1177/1091581810369565

Bbosa GS, Kitya D, Ogwal-Okeng J (2013) Aflatoxins metabolism, effects on epigenetic mechanisms and their role in carcinogenesis. SCIRP 5(10A):14–34. https://doi.org/10.4236/health.2013.510A1003

Behrens M, Hüwel S, Galla HJ, Humpf HU (2015) Blood-brain barrier effects of the *Fusarium* mycotoxins deoxynivalenol, 3 acetyldeoxynivalenol, and moniliformin and their transfer to the brain. PLoS One 10(11):e0143640. https://doi.org/10.1371/journal.pone.0143640

Burdock GA, Flamm WG (2000) Safety assessment of the mycotoxin cyclopiazonic acid. Int J Toxicol 19(3):195–218. https://doi.org/10.1080/10915810050074964

Byrem TM, Booren AM, Hill GM, Chu FS, Strasburg GM (1999a) The effect of cyclopiazonic acid on the development of pale, soft, and exudative pork from pigs of defined malignant hyperthermia genotype. J Anim Sci 77(1):166–172. https://doi.org/10.2527/1999.771166x

Byrem TM, Pestka JJ, Chu FS, Strasburg GM (1999b) Analysis and pharmacokinetics of cyclopiazonic acid in market weight pigs. J Anim Sci 77(1):173–179. https://doi.org/10.2527/1999.771173x

Callebaut F, Tangni EK, Debongnie P, Stals E, Huybrechts B, Waegeneers N, Delezie E, Van Cavret S, Lecoeur S (2006) Fusariotoxin transfer in animal. Food Chem Toxicol 44(3):444–453. https://doi.org/10.1016/j.fct.2005.08.021

Cavret S, Lecoeur S (2006) Fusariotoxin transfer in animal. Food Chem Toxicol 44(3):444–453. https://doi.org/10.1016/J.FCT.2005.08.021

Chen AJ, Jiao X, Hu Y, Lu X, Gao W (2015) Mycobiota and mycotoxins in traditional medicinal seeds from China. Toxins 7(10):3858–3875. https://doi.org/10.3390/toxins7103858

Corcuera LA, Vettorazzi A, Arbillaga L, Gonzalez-Peñas E, De Cerain AL (2012) An approach to the toxicity and toxicokinetics of aflatoxin B1 and ochratoxin A after simultaneous oral administration to fasted F344 rats. Food Chem Toxicol 50(10):3440–3446. https://doi.org/10.1016/j.fct.2012.06.048

Cui J, Xing L, Li Z, Wu S, Wang J, Liu J, Wang J, Yan X, Zhang X (2010) Ochratoxin A induces G2 phase arrest in human gastric epithelium GES-1 cells in vitro. Toxicol Lett 193(2):152–158. https://doi.org/10.1016/j.toxlet.2009.12.019

DellAfiora L, Dall'Asta C, Galaverna G (2018) Toxicodynamics of mycotoxins in the framework of food risk assessment—an *in silico* perspective. Toxins 10(2):52. https://doi.org/10.3390/toxins10020052

Deng J, Zhao L, Zhang NY, Karrow NA, Krumm CS, Qi DS, Sun LH (2018) Aflatoxin B1 metabolism: regulation by phase I and II metabolizing enzymes and chemoprotective agents. Mutat Res 778:79–89. https://doi.org/10.1016/j.mrrev.2018.10.002

Devreese M, Broekaert N, De Mil T, Fraeyman S, De Backer P, Croubels S (2014) Pilot toxicokinetic study and absolute oral bioavailability of the Fusarium mycotoxin enniatin B1 in pigs. Food Chem Toxicol 63:161–165. https://doi.org/10.1016/j.fct.2013.11.005

Duarte SC, Pena A, Lino CM (2009) Ochratoxin A non-conventional exposure sources a review. Microchem J 93(2):115–120. https://doi.org/10.1016/j.microc.2009.06.002

Effects in food-producing animals, chap 5 (n.d.). http://publications.iarc.fr/_publications/media/download/1375/23128bfd0453d69904d218c8aa9b9bf5480fc110.pdf

EFSA Panel on Contaminants in the Food Chain (CONTAM) (2014) Scientific Opinion on the risks to human and animal health related to the presence of beauvericin and enniatins in food and feed. Eur Food Safety Authority J 12(8):3802. https://doi.org/10.2903/j.efsa.2014.3802

El Khoury A, Atoui A, Ochratoxin A (2010) General overview and actual molecular status. Toxins 2(4):461–493. https://doi.org/10.3390/toxins2040461

Eriksen GS, Pettersson H, Lindberg JE (2003) Absorption, metabolism and excretion of 3-acetyl DON in pigs. Arch Anim Nutr 57(5):335–345. https://doi.org/10.1080/00039420310001607699

Escrivá L, Oueslati S, Font G, Manyes L (2017) *Alternaria* mycotoxins in food and feed: an overview. J Food Qual 2017. https://doi.org/10.1155/2017/1569748

Fleck SC, Pfeiffer E, Metzler M (2014) Permeation and metabolism of Alternaria mycotoxins with perylene quinone structure in cultured Caco-2 cells. Mycotoxin Res 30(1):17–23. https://doi.org/10.1007/s12550-013-0180-0

Flores-Flores ME, Lizarraga E, López de Cerain A, González-Penas E (2015) Presence of mycotoxins in animal milk: a review. Food Control 53:163–176

Fraeyman S, Devreese M, Broekaert N, De Mil T, Antonissen G, De Baere S, De Backer P, Rychlik M, Croubels S (2015) Quantitative determination of tenuazonic acid in pig and broiler chicken plasma by LC-MS/MS and its comparative toxicokinetics. J Agric Food Chem 63(38): 8560–8567. https://doi.org/10.1021/acs.jAfc.5b02828

Fraeyman S, Devreese M, Antonissen G, De Baere S, Rychlik M, Croubels S (2016) Comparative oral bioavailability, toxicokinetics, and biotransformation of enniatin B1 and enniatin B in broiler chickens. J Agric Food Chem 64(38):7259–7264. https://doi.org/10.1021/acs.jAfc.6b02913

Goyarts T, Dänicke S (2006) Bioavailability of the *Fusarium* toxin deoxynivalenol (DON) from naturally contaminated wheat for the pig. Toxicol Lett 163:171–182. https://doi.org/10.1016/j.toxlet.2005.10.007. PMID:16326049

Hadjeba-Medjdoub K, Tozlovanu M, Pfohl-Leszkowicz A, Frenette C, Paugh RJ, Manderville RA (2012) Structure-activity relationships imply different mechanisms of action for Ochratoxin A-mediated cytotoxicity and genotoxicity. Chem Res Toxicol 25(1):181–190. https://doi.org/10.1021/tx200406c

Haschek WM, Rousseaux CG, Wallig MA, Bolon B, Ochoa R (eds) (2013) Haschek and Rousseaux's in handbook of toxicologic pathology. Academic Press, Boston

Jestoi M, Rokka M, Järvenpää E, Peltonen K (2009) Determination of Fusarium mycotoxins beauvericin and enniatins (A, A1, B, B1) in eggs of laying hens using liquid chromatography–tandem mass spectrometry (LC–MS/MS). Food Chem 115(3):1120–1127. https://doi.org/10.1016/j.foodchem.2008.12.105

Jonsson M, Jestoi M, Nathanail AV, Kokkonen UM, Anttila M, Koivisto P, Karhunen P, Peltonen K (2013) Application of OECD guideline 423 in assessing the acute oral toxicity of moniliformin. Food Chem Toxicol 53:27–32. https://doi.org/10.1016/j.fct.2012.11.023

Kongkapan J, Poapolathep A, Owen H, Giorgi M (2015) A brief overview of our current understanding of nivalenol: a growing potential danger vet to be fully investigated. Israel J Vet Med 71:1–7

Kőszegi T, Poór M (2016) Ochratoxin A: molecular interactions, mechanisms of toxicity and prevention at the molecular level. Toxins 8(4):111. https://doi.org/10.3390/toxins8040111

Liu J, Wu S, Shen H, Cui J, Wang Y, Xing L, Wang J, Yan X, Zhang X (2015) Ochratoxin A induces DNA damage and G2 phase arrest in human esophageal epithelium Het-1A cells in vitro. J Toxicol Sci 40(5):657–665. https://doi.org/10.2131/jts.40.657

Malekinejad H, Akbari P, Allymehr M, Hobbenaghi R, Rezaie A (2011) Cyclopiazonic acid augments the hepatic and renal oxidative stress in broiler chicks. Hum Exp Toxicol 30(8): 910–919. https://doi.org/10.1177/0960327110384285

Malir F, Ostry V, Pfohl-Leszkowicz A, Toman J, Bazin I, Roubal T (2014) Transfer of ochratoxin A into tea and coffee beverages. Toxins 6(12):3438–3453. https://doi.org/10.3390/toxins6123438

Manderville RA (2005) A case for the genotoxicity of ochratoxin A by bioactivation and covalent DNA adduction. Chem Res Toxicol 18(7):1091–1097. https://doi.org/10.1021/tx050070p

Manderville R, Pfohl-Leszkowicz A (2008) Bioactivation and DNA adduction as a rationale for ochratoxin A carcinogenesis. World Mycotoxin J 1(3):357–367. https://doi.org/10.3920/WMJ2008.x039

National Toxicology Program (1989) Toxicology and carcinogenesis studies of ochratoxin A (CAS No. 303-47-9) in F344/N rats (gavage studies). National Toxicology Program technical report series, vol 358, pp 1–42

Norred WP (1990) Cyclopiazonic acid: toxicity and tissue distribution. Vet Hum Toxicol 32 (suppl):20–26

Norred WP, Morrissey RE, Riley RT, Cole RJ, Dorner JW (1985) Distribution, excretion and skeletal muscle effects of the mycotoxin [14C] cyclopiazonic acid in rats. Food Chem Toxicol 23(12):1069–1076. https://doi.org/10.1016/0278-6915(85)90054-7

Ohta M, Matsumoto H, Ishii K, Ueno Y (1978) Metabolism of trichothecene mycotoxins: II. Substrate specificity of microsomal deacetylation of trichothecenes. J Biochem 84(3): 697–706. https://doi.org/10.1093/oxfordjournals.jbchem.a132175

Ostry V, Malir F, Ruprich J (2013) Producers and important dietary sources of ochratoxin A and citrinin. Toxins 5(9):1574–1586. https://doi.org/10.3390/toxins5091574

Pahlke G, Tiessen C, Domnanich K, Kahle N, Groh IA, Schreck I, Weiss C, Marko D (2016) Impact of Alternaria toxins on CYP1A1 expression in different human tumor cells and relevance for genotoxicity. Toxicol Lett 240(1):93–104. https://doi.org/10.1016/j.toxlet.2015.10.003

Pavlović M, Plestina R, Krogh P (1979) Ochratoxin A contamination of foodstuffs in an area with Balkan (endemic) nephropathy. Acta Pathol Microbiol Scand B Microbiol 87(1–6):243–246. https://doi.org/10.1111/j.1699-0463.1979.tb02433.x

Pestka J (2010a) Toxicological mechanisms and potential health effects of deoxynivalenol and nivalenol. World Mycotoxin J 3(4):323–347. https://doi.org/10.3920/WMJ2010.1247

Pestka JJ (2010b) Deoxynivalenol: mechanisms of action, human exposure, and toxicological relevance. Arch Toxicol 84(9):663–679. https://doi.org/10.1007/s00204-010-0579-8

Pfohl-Leszkowicz A, Manderville RA (2012) An update on direct genotoxicity as a molecular mechanism of ochratoxin a carcinogenicity. Chem Res Toxicol 25(2):252–262. https://doi.org/10.1021/tx200430f

Poapolathep A, Sugita-Konishi Y, Doi K, Kumagai S (2003) The fates of trichothecene mycotoxins, nivalenol and fusarenon-X, in mice. Toxicon 41(8):1047–1054. https://doi.org/10.1016/S0041-0101(03)00089-8

Poapolathep A, Sugita-Konishi Y, Phitsanu T, Doi K, Kumagai S (2004) Placental and milk transmission of trichothecene mycotoxins, nivalenol and fusarenon-X, in mice. Toxicon 44(1):111–113. https://doi.org/10.1016/j.toxicon.2004.04.005

Poapolathep A, Poapolathep S, Sugita-Konishi Y, Imsilp K, Tassanawat T, Sinthusing C, Itoh Y, Kumagai S (2008) Fate of fusarenon-X in broilers and ducks. Poult Sci 87(8):1510–1515. https://doi.org/10.3382/ps.2008-00008

Pohland AE, Schuller PL, Steyn PS, Van Egmond HP (1982) Physicochemical data for some selected mycotoxins. Pure Appl Chem 54(11):2219–2284. https://doi.org/10.1351/pac198254112219

Prelusky DB, Rotter BA, Rotter RG (1994) Toxicology of mycotoxins. In: Miller JD, Trenholm HL (eds) Mycotoxins in grain: compounds other than aflatoxin. Eagan Press, St Paul, pp 359–404

Rushing BR, Selim MI (2019) Aflatoxin B1: a review on metabolism, toxicity, occurrence in food, occupational exposure, and detoxification methods. Food Chem Toxicol 124:81–100. https://doi.org/10.1016/j.fct.2018.11.047

Saengtienchai T, Poapolathep S, Isariyodom S, Ikenaka Y, Ishizuka M, Poapolathep A (2014) Toxicokinetics and tissue depletion of Fusarenon-X and its metabolite nivalenol in piglets. Food Chem Toxicol 66:307–312. https://doi.org/10.1016/j.fct.2014.01.053

Schelstraete W, Devreese M, Croubels S (2020) Comparative toxicokinetics of *Fusarium* mycotoxins in pigs and humans. Food Chem Toxicol 137:111140. https://doi.org/10.1016/j.fct.2020.111140

Schuchardt S, Ziemann C, Hansen T (2014) Combined toxicokinetic and in vivo genotoxicity study on Alternaria toxins. EFSA Support Publ 11(11):679E. https://doi.org/10.2903/sp.efsa.2014.EN-679

Schwartz GG (2002) Hypothesis: does ochratoxin A cause testicular cancer? Cancer Causes Control 13(1):91–100

Shim WB, Ha KS, Kim MG, Kim JS, Chung DH (2014) Evaluation of the transfer rate of ochratoxin A to decoctions of herbal medicines. Food Sci Biotechnol 23(6):2103–2108. https://doi.org/10.1007/s10068-014-0286-y

Springler A, Vrubel GJ, Mayer E, Schatzmayr G, Novak B (2016) Effect of Fusarium-derived metabolites on the barrier integrity of differentiated intestinal porcine epithelial cells (IPEC-J2). Toxins 8(11):345

Sugita-Konishi Y, Nakajima T (2009) Nivalenol: the mycology, occurrence, toxicology, analysis and regulation. In: Mycotoxins in food, feed and bioweapons. Springer, Berlin, pp 253–273

Trivedi AB, Doi E, Kitabatake N (1992) Detoxification of ochratoxin A on heating under acidic and alkaline conditions. Biosci Biotechnol Biochem 56(5):741–745. https://doi.org/10.1271/bbb.56.741

Turesky RJ (2005) Perspective: ochratoxin A is not a genotoxic carcinogen. Chem Res Toxicol 18(7):1082–1090. https://doi.org/10.1021/tx050076e

Van der Stegen GH, Essens PJ, Van der Lijn J (2001) Effect of roasting conditions on reduction of ochratoxin A in coffee. J Agric Food Chem 49(10):4713–4715. https://doi.org/10.1021/jf0105586

Veprikova Z, Zachariasova M, Dzuman Z, Zachariasova A, Fenclova M, Slavikova P, Vaclavikova M, Mastovska K, Hengst D, Hajslova J (2015) Mycotoxins in plant-based dietary supplements: hidden health risk for consumers. J Agric Food Chem 63(29):6633–6643. https://doi.org/10.1021/acs.jAfc.5b02105

Vidal A, Sanchis V, Ramos AJ, Marín S (2015) Thermal stability and kinetics of degradation of deoxynivalenol, deoxynivalenol conjugates and ochratoxin A during baking of wheat bakery products. Food Chem 178:276–286. https://doi.org/10.1016/j.foodchem.2015.01.098

WHO/FAO (2001) Ochratoxin A. In: Safety evaluation of certain mycotoxins in food, WHO food additives series, vol 47, pp 281–415

World Health Organization, International Agency for Research on Cancer (1993) Some naturally occurring substances: food items and constituents, heterocyclic aromatic amines and mycotoxins. IARC monographs on the evaluation of the carcinogenic risk of chemicals to humans, p 56

Wu Q, Dohnal V, Huang L, Kuča K, Yuan Z (2010) Metabolic pathways of trichothecenes. Drug Metab Rev 42(2):250–267. https://doi.org/10.3109/03602530903125807

Zhao Z, Liu N, Yang L, Deng Y, Wang J, Song S, Lin S, Wu A, Zhou Z, Hou J (2015) Multi-mycotoxin analysis of animal feed and animal-derived food using LC–MS/MS system with timed and highly selective reaction monitoring. Anal Bioanal Chem 407(24):7359–7368. https://doi.org/10.1007/s00216-015-8898-5